From Models to Simulations

This book analyses the impact computerization has had on contemporary science and explains the origins, technical nature and epistemological consequences of the current decisive interplay between technology and science: an intertwining of formalism, computation, data acquisition, data and visualization and how these factors have led to the spread of simulation models since the 1950s.

Using historical, comparative and interpretative case studies from a range of disciplines, with a particular emphasis on the case of plant studies, the author shows how and why computers, data treatment devices and programming languages have occasioned a gradual but irresistible and massive shift from mathematical models to computer simulations.

Franck Varenne is Associate Professor of philosophy of science at the University of Rouen (Normandy – France) and associate researcher at IHPST (CNRS – Paris). His research focuses on the history and epistemology of formal models and computer simulations in contemporary science, especially in biology and geography. He has published around fifty-five articles and chapters. He has also published eight books and co-edited three collective books.

History and Philosophy of Technoscience
Series Editor: Alfred Nordmann

Titles in this series

From Models to Simulations

Franck Varenne

Routledge
Taylor & Francis Group

LONDON AND NEW YORK

First published 2019
by Routledge
4 Park Square, Milton Park, Abingdon, Oxon OX14 4RN

and by Routledge

52 Vanderbilt Avenue, New York, NY 10017, USA

First issued in paperback 2020

This book is a translation and an update of the French book *Du modèle à la
simulation informatique*, Paris, Vrin, coll. "Mathesis", 2007. Translated by
Karen Turnbull. This publication is funded by MECS Institute for Advanced
Study on Media Cultures of Computer Simulation, Leuphana University
Lüneburg (German Research Foundation Project KFOR 1927).

British Library Cataloguing in Publication Data
A catalogue record for this book is available from the British Library

Library of Congress Cataloging in Publication Data
Names: Varenne, Franck, author.
Title: From models to simulations / Franck Varenne.
Description: Abingdon, Oxon ; New York, NY : Routledge, [2019] |
    Series: History and philosophy of technoscience | Includes
    bibliographical references and indexes.
Identifiers: LCCN 2018013718| ISBN 9781138065215 (hardback : alk. paper)
    | ISBN 9781315159904 (e-book)
Subjects: LCSH: Biological systems—Mathematical models. | Biological
    systems—Computer simulation.
Classification: LCC QH324.2 .V37 2019 | DDC 570.1/13—dc23
LC record available at htps://lccn.loc.gov/2018013718

ISBN 13: 978-0-367-58662-1 (pbk)
ISBN 13: 978-1-138-06521-5 (hbk)

DOI: 10.4324/9781315159904

Contents

Figures

Acknowledgments

This book is an extended and updated translation of the work first brought out in 2007 by the French publisher Vrin. Its release in English gives me the opportunity to thank, first and foremost, publishers Denis Arnaud (Vrin) and Robert Langham (Routledge), who backed this project, as well as Alfred Nordmann, who enthusiastically included this work in his collection. I also warmly thank the Institute of Advanced Study on Media Cultures for Computer Simulation (MECS) of the Leuphana University (Lüneburg, Germany) for helping me bring this project to fruition, first by inviting me to Lüneburg on a senior researcher fellowship in their extraordinary institute, and then helping to fund its translation costs. I would also like to warmly thank Sebastian Vehlken (MECS) who strongly supported this project. My warmest thanks go to Karen Turnbull, who, in close collaboration with me, has provided a remarkable piece of translation. Her language skills, as well as her knowledge of scientific and philosophical matters, helped her overcome the challenges inherent in the French version. If any ambiguities remain in this work, the responsibility is mine alone.

French abbreviations

ADEME   (*Agence de l'Environnement et de la Maîtrise de l'Énergie*) –
             French environment and energy management agency
AIP   (*Action Incitative Programmée*) – a type of INRA manage-
             ment policy aimed at setting up collaborations (both internal
             and with external laboratories) and stimulating funding for
             projects
AMAP   (*Atelier de Modélisation de l'Architecture des Plantes*) –
             Plant architecture modelling workshop
ATP   (*Action Thématique Programmée*) – Scheduled research
             initiative
bac+8   In the 1970s, French university students were required to
             submit two theses: a post-graduate thesis (known in French
             as the "*troisième cycle*" or "*bac[calaureat] plus 8 [years]*"),
             which is equivalent to the present-day PhD, and a State the-
             sis (called a "*thèse d'État*" or "*thèse d'habilitation*"), which
             was often written over a period of many years
Café, Cacao, Thé   (*Literally "Coffee, Cocoa, Tea"*) – IFCC Journal and also
             the name of an ORSTOM department
CIRAD   (*Centre de coopération internationale en recherche agronomique
             pour le développement*) – Agricultural Research Centre for
             International Development
CNRS   (*Centre nationale de la recherche scientifique*) – French
             National Centre for Scientific Research
DEA   (*Diplôme d'études approfondies*) – Diploma of advanced
             studies, comparable with a British Master's degree
DGRST   (*Délégation générale à la recherche scientifique et tech-
             nique*) – General delegation for scientific and technical
             research
EFPA   (*Écologie des forêts, prairies et milieux aquatiques*) – Forest,
             Prairie and Aquatic Environments
ENGREF   (*École nationale du génie rural, des eaux et des forêts*) –
             French National School of Forestry

ENSAT	(*École Nationale Supérieure d'Agronomie*) – French Higher National Engineering School of Agronomy – a competitive-entry engineering institution
ENST	(*École Nationale des Télécommunications*) – French National School of Telecommunications
EPHE	(*École Pratique des Hautes Études*) – Practical School of Higher Studies
EPIC	(*Établissement Public à Caractère Industriel et Commercial*) – Public-Sector Industrial and Commercial Enterprise – a type of public body established by statute in France
GERDAT	From 1980–1984: (*Groupement d'étude et de recherche pour le développement de l'agronomie tropicale*) – Study and Research Group for the Development of Tropical Agronomy
GERDAT	From 1985: (*Gestion de la Recherche Documentaire et Appui Technique*) – Management of Documentary Research and Technical Support (as of 1985, GERDAT became part of CIRAD, retaining the same acronym but with this new name)
IFCC	(*Institut Français du Café, du Cacao et autres plantes stimulantes*) – French Institute of Coffee, Cocoa and other Stimulant crops (later renamed "IRCC")
IN2P3	(*Institut national de physique nucléaire et de physique des particules*) – French National Institute of Nuclear Physics and Particle Physics
INAPG	(*Institut National Agronomique Paris-Grignon (INA P-G)*) – French National Agronomic Institute, Paris-Grignon
INRA	(*Institut national de recherche agronomique*) – French National Institute for Agricultural Research
INRIA	(*Institut national de recherche en informatique et automatique*) – French National Institute for Research in Computer Science and Automation, now known as the *Institut national de recherche dédié au numérique* – French National Institute for Computer Science and Applied Mathematics.
IRCC	(*Institut de Recherche sur le Café, le Cacao et autres plantes stimulantes*) – Institute for Research on Coffee, Cocoa and other Stimulant Crops (see IFCC above)
IRD	(*Institut de Recherche pour le Développement*) – Research Institute for Development, previously called ORSTOM
LHA	(*Laboratoire de l'Horloge Atomique*) – Atomic Clock Laboratory of the CNRS
LIAMA	(*Laboratoire franco-chinois d'informatique, d'automatique et de mathématiques appliquées*) – Franco-Chinese Laboratory of Informatics, Automation and Applied Mathematics
METALAU	(*METhode, Algorithmes et Logiciels pour l'AUtomatique*) – Method, algorithms and software for automation

ORSC	(*Office de la recherche scientifique coloniale*) – Office of Colonial Scientific Research
ORSTOM	(*Office de la recherche scientifique et technique outre-mer*) – Office of Overseas Scientific and Technical Research
PIAF	(*Physique et Physiologie Intégratives de l'Arbre en environnement Fluctuant*) – Integrative physics and physiology of trees in fluctuating environments
PNTS	(*Programme national de télédétection spatiale*) – French National Programme for Space-based Remote Sensing
SESA	(*Société de services et des systèmes informatiques et automatiques*) – Software and Engineering for Systems and Automata
SYRTE	(*Systèmes de Référence Temps Espace*) – Time and Space Reference Systems
ULP	(*Université Louis Pasteur*) – University of Strasbourg
UMR	(*Unité Mixte de Recherche*) – Joint Research Centre
USTL	(*Université des sciences et technologies du Languedoc*) – University of Science and Technology of Languedoc
UTC	(*Université de Technologie de Compiègne*) – University of Technology of Compiegne
X-ENGREF	Engineer from the *École Polytechnique* who has completed post-graduate practical training or internship (*école d'application*) at the French National School of Forestry (ENGREF: *École nationale du génie rural, des eaux et des forêts*)

Introduction

Many philosophical articles or books on computer simulation begin with general definitions or explanations, and then choose two or three specific sub-domains of science – along with a very small number of selected publications – that illustrate and confirm their definitions and interpretations. As a result, although they may be accurate regarding the epistemological meaning of a given technical solution, they sometimes lack a certain sensitivity to real field solutions, to their multiplicity and to the dramatic epistemological innovations that emerge mainly from the field. Other books on history or sociology of science may be more aware of both the diversity of technical solutions and the importance of field innovations. However, since a significant number of these books are multi-author volumes, they simply juxtapose, or at best loosely compare, the many descriptions of different technical and epistemological solutions, and their comparisons are made between simulations of different target systems with overly disparate formalisms, and methodological and computational solutions that are too heterogeneous. For this reason, although these publications may be particularly informative, most are not ultimately conclusive from an epistemological standpoint. Nor can they guard us against a sense of general dissonance. With such approaches, the meaning of the term "simulation", or even its understandable polysemy, remains vague and somewhat disheartening.

Exceptions to these two frequent limitations of the current literature on computer simulations can be found in some works regarding simulation techniques in a specific domain of objects whose evolution is studied across a sufficient lapse of historical time. A brilliant example exists in nuclear physics, namely the work of Peter Galison.[1] Since the early 2000s, however, it has become clear that there is often a greater diversity of simulation techniques and consequently of epistemological innovation in the biological and social sciences – which are constantly developing new computer simulations – than there is in physics, in contrast to the general rule in the technosciences in the immediate post-war period. This book can be seen as an attempt to help fill the gap in this regard.

Starting from the undeniable achievements, as well as from the limitations, of the previous studies of many other researchers, and based on a longitudinal case study in quantitative, mathematical and computational biology, this book first adopts a historical and comparative approach to the different research programmes operating in the same field: *modelling the growth and morphogenesis of single*

DOI: 10.4324/9781315159904-1

vegetative plants in botany, forestry and agronomy. Having chosen this relatively vast domain, along with these three different types of approach, and without neglecting the personal, social and institutional factors, this book's methodical approach is mainly based on an intellectual and comparative analysis of the different solutions to modelling and simulation issues both in the theoretical approaches to plant growth and in the more applied and technoscientific approaches that have emerged since the 1950s.

It is important to note that the content of this book is based not only on analyses and comparisons of publications (in English, German and French), but also on more than twenty interviews or personal correspondence with some of the key actors. Using a diachronic and comparative perspective, the book describes the exact field involved, as well as the technical and formal reasons and the epistemological decisions that explain why each kind of computer simulation of the various aspects of plants gradually replaced the mathematical models, i.e., the pre-existing models that originated from theoretical biology, biometry or morphometrics. It is hoped that, as a consequence, this book will give the reader the epistemological and conceptual acuity that seems necessary today to avoid many interpretative confusions: namely the confusions between quantification and formal modelling, between laws and models, between models and simulations, between mathematical models, computational models and simulation models, between simulations of models and models of simulations, and, last but not least, between different types of computer simulations.

With the aim of presenting an updated and extended version, supplemented with more in-depth epistemological insights, this English translation includes several additions to the original introduction and conclusion, as well as to a number of the chapters. Chapter 8, entitled "Twenty-one functions of models and three types of simulation – classifications and applications", is entirely new, however. This chapter's aim is first of all to present a distinctive general classification of the epistemic functions of scientific models, as well as a classification of the different types of computer simulation. This approach is intended to remain very general in scope, in the hope that it will thus benefit research on models and simulations in completely different fields from those of plants or biology. Its content is the result of a work of comparison and induction carried out not only on the basis of the comparative history of plant models presented herein, but also on the basis of several collaborative research efforts that have been carried out since then, as well as on my own, even more recent, large-scale research in the field of comparative history of models and simulations in geography.[2] Next, with the two-fold aim of confirming the relevance of these conceptual analyses based on the available evidence on the one hand, and, on the other, of reviewing the comparative history recounted in Chapters 1 to 7 from a more discerning and discriminating epistemological perspective, Chapter 8 will end with a systematic application of these classifications to the different types of models and simulations encountered in the case of plants.

It may be remarked that a fairly substantial portion of this book focuses on French research and researchers, leading to the conclusion that this reflects an unjustified bias. With regard to plant studies, however, there are certain situations specific to

France, such as the enduring existence of French research institutions in previously colonized tropical countries such as Côte d'Ivoire (see Chapter 3), even long after these countries obtained political independence. Such situations played a large part in the dynamics and focus of the research reported here insofar as they enabled quantitative botanists to have very early and direct access to the huge diversity of tropical flora, while, at the same time, providing them with access to adequate instrumentation. As a counterbalance to what may potentially be perceived as a French-oriented bias, however, I also describe in detail, in Chapter 1, how and why Jack B. Fisher, together with Hisao Honda, were among the first botanists to attempt to tackle the problem of using computers to faithfully represent the growth and architecture of vegetative plants. I also explain why, as a perhaps too rigorous botanist, Fisher ultimately decided not to develop his simulations further. It is no accident that Fisher also worked in a quasi-tropical context, in the Fairchild Tropical Garden of Miami; like the researchers in Côte d'Ivoire, he was also exposed to the incentive of maximal diversity. The fact remains, however, that for a long time, apart from some tropicalists such as Fisher, most of the researchers in quantitative botany and forestry working in North America and Great Britain remained in the mainstream of classical mathematical modelling. Important exceptions can be found in Canada, in the Prusinkiewicz school in particular, and also – from the 1990s onwards – in Germany and Finland. I have also been careful to include these exceptions and their specific "pre-histories" in Chapters 2 and 7 in particular.

The period of history involved here covers the end of the 1960s up to the first few years of the 21st century. This period is, of course, not without antecedents. This work does not aim to dwell in detail on the periods that preceded it, but, in order to better understand its specific technical and epistemological aspects, and especially what I propose to identify as a transition "from mathematical model to software-based simulation", I consider it necessary to give a preliminary outline in this introduction of the way in which the formal models took root and were originally grasped and used in the study of plant growth.[3]

Thus, when we examine the period prior to the one we will study – the period from the 1920s to the beginning of the 1960s – we can see two different epochs emerge fairly clearly. The first corresponds to the years before the spread of the digital computer. It extends from the 1920s to the end of the 1940s. In those years, mathematical modelling permeated several sectors of biology. Briefly put, it had a two-fold effect of increasing the available types of formalisms and of diversifying the epistemic functions of the mathematical formalizations, in contrast to the usual functions attributed to the mathematical laws and theories traditionally used in biology. A second, much shorter epoch stretched from the beginning of the 1950s to the mid-1960s. This was the epoch of the first impacts of computerization on formal models, and it included, in particular, the appearance of the first computer simulation techniques. These techniques would interfere in both a competing and a constructive manner with the formal modelling practices that were then still flourishing. Over the next few paragraphs I will describe these two epochs in somewhat greater detail.

With regard to the first epoch, what the scientists often called the "formal-model method" became progressively established in the quantitative biology of morphogenesis, based on four different areas: biometry; population biology; mathematical biology; and biocybernetics. The formal-model method, in biometry in particular, had its roots in the epistemological decision of Ronald A. Fisher[4] to abandon the Bayesian interpretation of statistics and instead to propose – in line with the theory of errors that had emerged from works on astronomical observation, and in the wake of the famous article by Student[5] on probable error in the estimation of a mean – a "hypothetical law" for estimating statistical parameters in the case of small sample sizes. In my view, this hypothetical law, which was explicitly free of any attempt at representation and thus of rootedness in the actual causal connections, acted as the first detached formalization, or first formal model in the full sense of the term as it is used in biology. The "hypothetical law" itself took the form of a frame of reference for field data, and was widely termed "model" from the end of the 1940s. It was this fictive and detached formalization that would to a large extent serve as a prototype for the other types of formal model in biology, including in population biology, from the 1920s onward. This first epoch may be called the "epoch of detachment of formalisms", since it is characterized by an increasing and normalized use of this type of formal construct, known as formal models.

In this context, by "formal model" I am referring to any type of formal construct of a logical or mathematical format with an axiomatic homogeneity that is capable of answering certain questions and fulfilling certain functions (cognitive, empirical, communication-related) with respect to an object, a system or an observable phenomenon. The formal model differs from theory in its validity, which is often only local, in its prior adaptation to certain questions that are posed at the outset, and in its inability to directly produce general results in the form of theorems. It should be pointed out already here that it also differs from simulation, although the term "simulation" is ambiguous, since it designates both a symbol-processing operation and the symbolic result of that processing. I will revisit all these points in greater detail in Chapter 8. We could say that, as a first approach, a computer simulation – insofar as it is a process – may be seen as a computer-assisted symbolization and formalization technique consisting of two distinct steps. During the first step, termed *operative*, symbols that more or less realistically represent elements of an actual or fictive target system interact step by step in accordance with rules, and these rules themselves may represent certain real or fictive mechanisms of the target system. Adopting a term used in connectionist artificial intelligence, I consider that this step is based on a *sub-symbolic*[6] use of certain formalisms and certain systems of symbols. The second step of a simulation consists of an equally symbolic processing of the results of the first step. This step may be described as *observational*, and it consists of a set of *reckonings, measurements, observations* or *visualizations* regarding the outcome of the first step. The main epistemic function of the computer simulations that were initially the most widespread, i.e., "numerical simulations", was to replace impossible formal calculations with *measurements carried out on these interaction results*. Thus, these interactions did indeed take place between symbols that had also been given the status of sub-symbol: they were sub-symbols from the point of view of the resulting patterns. Not all the computer simulations still have the primary function of replacing an analytically intractable calculation, but all retain this two-step structuration. One consequence of this structuration is that, as we will see in this historical and comparative case study, even though a computer simulation uses formal models, unlike those models it is not always a homogeneous formal construct. For that reason, a simulation does not necessarily have to be based on a single selective viewpoint on the target system, and nor is it obliged to have a formal homogeneity because of its format. This point will form one of the main established facts of this investigation, and I will return to it in more detail, giving specific examples.

First, let us return to the characterization of the formal-model method in the empirical sciences, and to its innovative nature in biology in the first epoch, starting in the 1920s. It should be noted that – although mechanical models, in the sense still used by William Thomson (Lord Kelvin), Maxwell or Boltzmann in the 19th century, responded to a demand for *visualization of calculations, or picturability*, and although a formal model in that part of mathematics known as the mathematical theory of models was itself still considered to be a more concrete, albeit mathematical, representation of a purely formal theory – it was no longer the model's concrete and representable nature or its ability to interpret a theory that were sought from the 1920s onwards in the "formal-model method". Instead, what was sought was an ability to directly and formally represent certain relationships between observable properties or physical quantities, if necessary in a way that remained purely phenomenological, i.e., precisely without representing a credible underlying mechanism, but also without interpreting a formal theory that had been explained in advance. As a result, the *formal model* tended to be a direct formalization that was no longer based exclusively either on a prior physical model or on a more abstract theory. This type of epistemic function was new for models in the empirical sciences. By virtue of the model's henceforth formal nature, and owing to this new function it was given, the model may seem to conflict with the nature and the epistemic functions of the traditional formal laws. Nevertheless, it was still recognized as a model and not as a law: these two characteristics – its local validity and the fact that the justification for its construction is based on a particular question and a precise perspective – are still used to differentiate between the scope and function specific to a formal model and those specific to a mathematical law.

Thus, because of the specific different epistemic functions now given to models, and because of the correlative divergences in terms of fieldwork epistemology, this first epoch, which saw the emergence of the model method, is characterized by a general renewal of the legitimization of formalisms in the life sciences, in particular for studying morphogenesis: the formalisms became more varied once they were no longer necessarily determined by representations resulting solely from physics or theories that could be completely and formally mathematized. The legitimization of these formalisms could itself become more varied. The term "model"[7] becomes more frequent in the literature, and then systematic. In this, mathematics first played a purely pragmatic role as a tool for data investigation or data representation, combined with an epistemology favouring detached formalizations and pluralism, an epistemology that was in fact often fictionalistic and instrumentalistic. Parallel to this effect favouring a pluralistic epistemology of formal models in applied biology, mathematics still played a major role for the most speculative of the bio-mathematicians in their theoretical-mathematical models. This was no longer a role of symbolic replication of entities and elementary mechanisms (entity realism), however, but rather a role as a means of revealing the directly mathematical-type stable structures (structural realism). It is the recognition of the latter epistemic function that emerges, for example, in the transition from biophysics to biotopology that the biomathematician Nicholas Rashevsky first invoked in 1953.[8] Thus, not only with regard to the rise of descriptive mathematics in applied biology, but also with respect to the most theoretical works, mathematical ingenuity was directed at what I have called a *detachment of formalisms*. To that extent, this mathematical ingenuity would partially replace the models and metaphors that traditionally derived from physics and its related disciplines.

Let us turn now to the second epoch, which precedes our own and extends from the end of the 1940s to the early 1960s. This epoch has a series of features that I will sum up briefly. First of all, owing to the availability of digital computers, *digital simulation* developed very early on alongside the formal models, but in an equally polymorphic manner. A different hurdle was cleared from the path of formal modelling for plant morphogenesis with each of the contributions from the new authors – all of whom were mathematicians and not biologists. Alan Turing (1952) emphasized the contribution of the discretization of formalisms. Murray Eden (1960) highlighted the need to formalize real random events with simulated random events through "stochastic" models based on the laws of probability. Lastly, Stanislaw Ulam (1962) demonstrated the importance of the spatialization of formalisms in order to formalize spatial phenomena.[9] This would mark the beginning of cellular automata. In this context, computer simulation proved from the start to be a formalization strategy that operated on a lower level of abstraction than classic formal modelling, with a complicated manual processing that was offset by a massively iterative processing delegated to the machine. It is in fact in this sense that computer simulation relies on a sub-symbolic use of the usual sets of axioms. The formal models in a computer simulation are not calculated formally by the computer thanks to deductive rules; instead it is their axiomatic functioning that is simulated by the sub-symbolic representations, which in turn possess their own set of rules and axioms. These other sets of rules and axioms are at times – but not always – of a more immediately interpretable nature, as is the case, for example, of discrete representations that use one-to-one relations between single-memory addresses in the computer and neutrons in the first computerized nuclear physics.

Until the beginning of the 1960s, however, each of these digital simulations of growing living beings, by selectively sub-symbolizing a formal representation, extended the power of expression of the formal model in accordance with a maximum of one or two dimensions that had until then been inaccessible to mathematics. Each of these simulations thus gave rise to just one selective digital representation. Moreover, none of these simulations could be fed precise field data, which would have enabled an effective calibration to be carried out. All of the results of these digital simulation processes thus remained merely qualitative models. From this point of view, these simulation processes were ultimately comparable to the contemporaneous theoretical-mathematical models that, in their turn, still sought to explain by invoking a single fundamental or predominant mechanism, such as those of biophysics, biotopology, relational biology, differential topology or plant-structure thermodynamics.[10] In this context, a computer that simulates remains an unrefined and purely qualitative simulator. It produces graphs or curves that admittedly bring to mind the shapes found in nature. But this similarity remains purely qualitative. Thus, whether it is a case of formal models or of those first computer simulations, formal multiplicity and diverging formal solutions remain the rule. It is the divergence and dispersion of mere intention, of speculation and of selective mathematical actions without a grip on the world of real plants.

As for field modelling, such as the modelling used during this second epoch in agronomy and forestry – the very modelling that was most expected to have a grasp on reality – formal divergence and diversity were in fact its method, its *credo*. For multifactor experimental designs applied to increase in biomass, for improvements in crop management, for problems of blight control, the biometric models of plant growth worked very well. They were designed to do so. Nonetheless, despite all this newly available formal diversity, these models failed when it came to focusing on monitoring of morphogenesis on the scale of the individual plant. As a result of this failure, they ultimately rather glaringly revealed the unavoidably perspectivist and selective nature of the formal model. The problem was, so they said, that the properties of a living organism could not all be formalized *at the same time*. But what may seem here to be a defect of the model, field biometrics often decides to interpret as a quality of nature, as proof that we are indeed dealing with nature, in its infinite complexity. These formal field models are selective in their perspective. What is more, they are mutually exclusive: nothing could be more normal, as we often read in the scientific literature itself, than this fruitful tension between representation and action. Beyond certain cultural differences, an epistemology of a pragmatic type that is adopted principally in the English-speaking countries due to the overwhelming influence of nominalism and of pragmatist philosophies may, strangely enough but very significantly, harmonize on certain points with a dialectic-type epistemology that is more specifically adopted on the European continent, and in particular in France, due to the persistent influence in this context of Hegelian rationalism and dialectic materialism. During this epoch, these two epistemologies, which were otherwise so distinct, could thus be seen to confirm each other's intuitions, since both claimed that it was necessary to renounce the aim of simultaneously representing the infinite multiplicity of dimensions of the object under study. Both claimed that it was necessary to try to offset this impossibility by a multiplicity of formal and selective modelling approaches to that object.[11] It is true that these formal approaches remain mutually incompatible, because they are axiomatically not co-calculable. As a result, they can only be juxtaposed but not aggregated. We may pass from one to the other, but they are never aggregated with each other.

As for mathematical models with a theoretical function, those who are dedicated to these models in theoretical biomathematics may lament their diversity, while at the same time nonetheless also contributing to increasing this diversity. Thus they seek to make them not exclusive, but rather mutually absorbing, since in that way they can demonstrate that they are capitalizing on earlier works and that they are doing better than them. The metaphor I am suggesting here is that of absorption: this is the direct opposite of the metaphor of aggregation that applies for integrative simulations. I would say that a theoretical-mathematical model is *absorbent* because it is conceived to replace and emulate one or several other models, while at the same time bringing its own epistemic contribution. It emulates other models in the sense that it seeks to be more general by dispensing with the explicit formulation of the preceding theoretical-mathematical model, but fulfilling almost the same epistemic functions of comprehension – and sometimes of partial prediction – as the previous one while adding several other functions of its own. From this point of view, the formalisms of theoretical biology are in competition with each other for theoretical dominance. They neither accept nor seek a peaceful juxtaposition. They seek to reduce each other in the secret hope that there will remain only one at the end: this is the process of absorption. But any contemporary historian of science can nonetheless see that biomathematics fails to propose a final, convincing absorption, namely a comprehensive general theory of morphogenesis and growth that would be based, for example, on information, entropy, the mathematical theory of catastrophes, on fractals, or indeed on a general theory of signals or networks. The result of this relative failure is that, rather ironically, and even tragically from their point of view, these theoretical models actually become very different also in the scientific literature. In these multiple works of resistance to multiplicity, to perspectivist and pragmatist modelling, as well as to the dispersion of detached field models, the search for a unique and monoformalized theoretical model – i.e., one that is formalized in only one sole mathematical set of axioms – plays the role of substitute for the lost and seemingly direct rooting of the old models in the physical world.

This second epoch therefore is characterized, on the one side, by a calm acceptance of the mutual incompatibility of models as long as they promote human action, and on the other by an uneasy rejection of that dispersion because it heralds a loss of meaning, in particular for those who disagree with pragmatism or dialectic rationalism. This, then, is the portrait of an epoch that, for other equally fundamental reasons (such as the changing social demands with regard to science in the post-war period, the recognized limitations of the capabilities of instruments and formalisms, the changing objects of study), with relative coherence developed its own consensual epistemology of the plurality and dispersion of representations, ending with its later explicit affirmation during the 1980s in some research work and symposia on epistemology and science studies. In some ways, the movement towards a pluralization and dispersion of models that was specific to this second epoch is the same as what we are still witnessing today, in a large part of contemporary science. The epistemologies of the dispersion and disunity of science were able to come into being and find ways of justifying themselves during that epoch, in particular by exploiting the method of formal models and of correlative iconoclasm, or rejection of integral representation.

And yet, in the case of an object that is complex, because it is particularly composite, such as the plant for example, it turns out that monoformalized models, even when multiplied, or even when they have a statistical nature and only a pragmatic aim, are no more capable than monoformalized theories of providing predictive and effectively operational formalizations. And in the face of social demand, science has thus had to try to advance further still and circumvent this hurdle. This is essentially the reason why, as I will show more particularly in this work, from the mid-1970s onwards, botanists, agronomists, foresters and other plant specialists all turned towards *integrative software-based computer simulation*[12] based on an individual-based approach, since, thanks to the visualization devices and object-based computer languages, such simulation permitted the convergence of perspectives, scales and mechanisms, and thus of multiple formalisms. I will demonstrate that *software-based simulation* thus brought about two fundamental innovations. First, simulation broke with the supremacy of formal models and their associated epistemologies, albeit without downgrading them entirely. Next, it broke with the numerical simulation that had emerged in the immediate post-war period and that was still dependent on mathematical models and the assistance they provided. Indeed, software-based simulation made it possible to achieve precise calibration and, in many cases, quantitative prediction, or even – which remains a heresy for many – an outright "experiment on simulation", also known as a "virtual experiment". Its essential principle, as we will see, is what I propose to call *pluri-formalization* or, in other words, a computer integration of formalisms of different natures (logical, mathematical) and from different points of view. This latest-generation simulation, far from being simply a discretization of models, takes a position that is at times in competition with models and mathematics, insofar as it makes something that is not compatible mathematically compatible on the level of the programming language and of the computer program. This, to my mind, seems to be its most decisive contribution since the beginning of the 1990s. Its truly empirical nature obviously remains in question, and we will see this in detail during the investigation, and also in the conclusion, in the form of a comparative table. But the questions that arise in the matter of its empirical nature are in fact not all the same as those that have already arisen regarding the empirical nature of numerical simulation. I can already say that the formalization that such an integrative simulation carries out takes on a compactness and a depth due to the fact that several different perspectives, and therefore the approaches of several different disciplines (physiology, mechanics, architecture, etc.), are possible at the same time. Simulation thus breaks, at the very least, with the perspectivist and purely pragmatist epistemology that often accompanied the first formal models: modelling from a precise perspective, and with a precise objective. Software-based and object-based simulations go beyond an *integrative pluralism*[13] as well as a *selective realism*,[14] and truly implement an *integrated plurality*. It is thus a very different epistemic practice than traditional formal modelling and its technical extensions. It is therefore necessary to try to look at it in a different manner.

Although simulation was conceived from the practices of modelling, it has admittedly not made modelling disappear. But it has shifted, amplified and somewhat disaggregated modelling by giving a new status to the formalisms: a quasi-empirical status. It is here that lies the central role of the computer, the half-material, half-formal instrument that has contributed to building bridges of various types between the practices of minimally abstracted replication and the more classic practices of abstraction and calculation. Computer simulation was developed first of all in the form of so-called *numerical* simulation. In this form, it first served to resolve the mathematical models that were otherwise intractable, and in so doing made it possible to considerably extend the methods of calculation by finite elements that date back to mid-19th-century techniques for the calculation of structures. Since the 1990s, however, computer simulation has decisively broken with the monopoly of that single function of approximate calculation of models. At times, it even precedes the model. To such an extent that, for the past ten or fifteen years, far from limiting itself to the numerical resolution of mathematical models that have been conceived beforehand with one single set of axioms and from one single perspective, more and more scientists seek formal models on virtual integrative mock-ups or on pluriformalized integrative models of simulation. In such simulations, it is not just various homogeneous algorithmic rules that replace the mathematical laws (this is the case of the *algorithmic simulations* developed since the beginning of the 1960s), but these rules may go so far as to be fundamentally pluralistic, evolutive, heterogeneous and spread out over the different times and spaces of the computation. The order of priority between model and simulation is thus inversed: we simulate before we model. Software-based computer simulation thus is distinguished not just from numerical simulation, but also from algorithmic simulation.[15] Having now become the complex double of a reality that is perceived and conceived as complex, computer simulation has ended up melding with the experimentation per se and the monoformalized modelling. Thus, since becoming software-based in the 1970s, simulations have had a tendency to become considerably more complex. They now allow an *integrative and figurative realism*, and these detailed, multiscale and multi-process representations have taken on an altogether remarkable weight. In return, when they are validly calibrated and stabilized, these simulation strategies make it possible for modellers to leave behind the completely simulated approaches and to enter into a phase of formalization that, starting in Chapter 7, I propose to call *remathematization*. Thus, it becomes more and more clear that in certain domains that study objects, such as plants, that are considered to be complex, searching for a formal model directly from the data, without prior integrative simulation, now seems to be truly too arbitrary and something that should be avoided. Today, a mathematical modelling that aims to skip the step of integrative simulation, even if its declared aim is merely theoretical, heuristic or pragmatic, becomes more and more open to question. Thus this inversion of priority between the practice of simulation and the practice of mathematical formalization is not the least of the recent contributions of computerization in the sciences that use models.

What particular technical and epistemological choices determined this type of decisive innovation? What are the precise types of the various integrations and convergences that, after a period of detachment and then of pluralization and dispersion of formal models, characterize this new epoch into which we have entered – an epoch in which, as we will see, plant-growth models and simulations have been precursors to a considerable extent?

Might it not be said – with regard to the formalisms that are applied to the objects studied by the empirical sciences – that this epoch of integration and convergence of formalisms in fact testifies to a simple practice of "rerooting"? In other words: to what extent can it be said that the convergences made possible by computerizing the methods of formalization exhibit neither a return back towards a mathematicist essentialism according to which the world is seemingly written in a single mathematical language, nor an escape forwards to a naïve and illusory figurative realism, the result of our apparent fascination with images and virtual worlds, rather than a desire for comprehension and true science? For that matter, in what sense can it be said of a computer simulation that it possesses an empirical dimension? Is this true of all simulations? Otherwise, of which ones is this true, and why? What are the limitations of the knowledge conferred by software-based and object-based simulations if we are already able to perceive them? What precise epistemological lessons can we already draw from this very recent evolution? And finally, what new conceptual and terminological propositions can the modern epistemology of models and simulations adopt to try to go a step further than the old epistemologies of models that, in the 20th century, were successively or concurrently of syntactic (logicism), dialectic, semantic and then pragmatic influence?[16]

This historical and interpretative investigation, which I have the honour to submit here in updated form for English-speaking readers, attempts to answer some of these questions. It does so by choosing to focus on certain scientific works that have, to my mind, played a large part in determining this recent transition from model to simulation. As we will see, I have paid particular attention not only to the technical choices of these works, but also to the methodological and epistemological decisions that accompanied them, as well as, when necessary, to the administrative and institutional contexts that witnessed their emergence. This work, inspired by the reflections that cropped up during my own use of mathematical modelling and numerical simulation in the field of applied atomic physics,[17] is based primarily on field-survey work, on a systematic collection and analysis of publications and archives, and on oral and written interviews carried out with some twenty-odd of the main protagonists of this story. It is also based on the interpretation and epistemological contextualization of the various recent schools and practices of modelling and simulation. Based on the idea that a philosophy of science cannot do without a history of science that is both very contemporary and highly comparative, this work aims to draw an epistemological lesson that is, if possible, enriched and differentiated regarding the different practices of formalization used in the empirical sciences – practices that have continued without cease to characterize modern science since its first great successes of the 17th century.

Notes

1 P. Galison, *Image and Logic*, Chicago: University of Chicago Press, 1997.
2 F. Varenne, *Théories et modèles en sciences humaines. Le cas de la géographie* [Theories and models in human sciences. The case of geography], Paris: Éditions Matériologiques, 2017.
3 The comparative history of this earlier period was the focus of another book, which has not yet been translated: F. Varenne, *Formaliser le vivant: lois, théories, modèles?* [Formalizing living beings: laws, theories, models?], Paris: Hermann, 2010.

4 R.A. Fisher, "Studies in crop variation, I. An examination of the yield of dressed grain from Broadbalk", *Journal of Agricultural Sciences*, 1921, 11, pp. 107–135.
5 W.S. Gosset (alias "Student"), "The probable error of a mean", *Biometrika*, 1908, 6, pp. 1–25.
6 See glossary.
7 Regarding the polysemy of this term, see glossary.
8 F. Varenne, "Nicholas Rashevsky (1899–1972): de la biophysique à la biotopologie" [Nicolas Rashevsky, (1899–1972): from biophysics to biotopology], *Cahiers d'Histoire et de Philosophie des Sciences*, Special Edition, 2006, pp. 162–163.
9 For a comparative analysis of these three contributions, see F. Varenne, *Formaliser le vivant . . .*, 2010, op. cit., partie III "La naissance des simulations" [Part III "The birth of simulations"], pp. 163–217.
10 On these various theoretical approaches, see F. Varenne, *Formaliser le vivant. . .*, 2010, op. cit., partie IV "Le tournant mathématiste des théories" [Part IV "The mathematicist turning point of theories"], pp. 219–275.
11 This mutual exclusion of multiple models is not necessary, however, suggests A.F. Schmid in *L'âge de l'épistémologie* [The age of epistemology], Paris: Kimè, 1998.
12 "Integrative" in no way signifies "integral".
13 S.D. Mitchell, *Biological Complexity and Integrative Pluralism*, Cambridge: Cambridge University Press, 2003.
14 P. Humphreys, *Extending Ourselves: Computational Science, Empiricism and Scientific Method*, Oxford: Oxford University Press, 2004.
15 See the terminological distinctions set out in Chapter 8 and the glossary.
16 For confirmation of this reading, see M.S. Morgan, M. Morrison (Eds), *Models As Mediators*, Cambridge: Cambridge University Press, 1999. For a debate on interpretation, see F. Varenne, *Les notions de métaphore et analogie dans les épistémologies des modèles et des simulations* [The concepts of metaphor and analogy in the epistemologies of models and simulations], Paris: Pétra, 2006.
17 Between 1993 and 1996 I was first a trainee Engineer and then Research Engineer at the Laboratoire de l'Horloge Atomique (LHA – Atomic Clock Laboratory) of the CNRS (Centre national pour la recherche scientifique – National Centre for Scientific Research), at Orsay, near Paris, during two periods covering a total of 15 months. This laboratory has since merged with the SYRTE laboratory. The SYRTE department – Systèmes de Référence Temps Espace (Time and Space Reference Systems) – belongs to the Paris Observatory – Paris Sorbonne Lettres Research University and is also associated with the CNRS – National Research Centre and University Pierre & Marie Curie (Paris 6) – Sorbonne University. Website: https://syrte.obspm.fr. I would like to take this opportunity to thank the colleagues I had the pleasure of working with then, and with whom I continued my training in physics and modelling: Pierre Cérez, Noël Dimarcq and Bertrand Boussert.

1   Geometric and botanic simulation

Until the early 1960s, the computer simulations used in morphogenesis problems had developed in an inchoate and divergent manner, based on issues that were essentially specific to professional mathematicians. It is important, therefore, to first reconstruct the factors that enabled these simulations to evolve towards greater biological realism. The simulations I will consider are of three types: geometry and probability-based, logic-based and pluriformalized. We will see how their implementation broke not only with the traditional uses of the computer as a calculator of models, but also with the pragmatic epistemologies of these models. These simulations, which were successfully adopted by biologists, were the first to establish closer links with work in the field, but they would not be the last. This series of initial intersections with the empirical opened the way to an era of convergences with many different dimensions. In this chapter, in particular, I will show that certain biologists (such as Dan Cohen and Jack B. Fisher) were ultimately able to make good use of simulation once they managed to accentuate its ability to produce a representation on a geometrical level – albeit at the cost of reducing their ability to take the temporal heterogeneity of plant-growth rules into consideration.

The probabilistic simulation of branching biological shapes: Cohen (1966)

We find the first use in biology of a particular type of discretized computer simulation in the work of the Israeli biologist Dan Cohen (born 1930). What interested Cohen (at The Hebrew University of Jerusalem) above all was the possibility of forming a theoretical argument on the processes of morphogenesis. His aim was to improve on the work of his MIT colleague Murray Eden (born 1920), by trying to find certain concepts that were specific to the botanical morphology and embryology of the time, including those of Conrad Hal Waddington (1905–1975), an embryologist and organicist. Eden, an engineer and mathematician, had worked with the linguist Morris Halle (born 1923) in 1959 on a letter-by-letter modelling of cursive writing, using a succession of elementary probabilistic choices with multiple branches. He demonstrated that a numerical simulation of this combinatorial analysis to biological morphology situation, when depicted on a plane (random branching on a grid of square cells), could, in a first approximation, be considered analogous to cellular multiplication.

DOI: 10.4324/9781315159904-2

In 1966 Cohen recognized the value of this spatial representation of intertwined calculations. He set out to theoretically test Waddington's hypothesis of a morphogenesis conceived as the result of a "hierarchically ordered set of interactions between genes, gene products and the external environment".[1] Cohen adopted the concept of "epigenetic landscape" that had been introduced by Waddington in 1957 in *The Strategy of the Genes*, in which he adapted to an ontogenetic scale the earlier (1932) notion of "adaptive landscape" proposed by the American geneticist Sewall Wright (1889–1988). Cohen's aim was to thereby link his own evolutionary ecology questions to this morphogenetic problem of epigenesis. Without wishing to precisely calibrate his morphogenesis model on actual cases, he nonetheless wished to demonstrate a general overall feasibility: the feasibility for living beings of undergoing progressive and adaptive growth and development without requiring reference to excessively complex laws, with such complexity being expressed in terms of "information". According to Cohen, if one could write a "minimal" program using the "simplest possible"[2] rules of generation for branching shapes that were already relatively realistic from a global and qualitative point of view (as judged by eye), then one could consider that the plausibility of Waddington's epigenetic hypothesis had been increased, along with the analogous hypotheses of evolutionary ecology.

To this end, a branching structure was drawn by a computer connected to a plotter. Cohen then followed Eden's probabilistic approach for branching. But in order to clearly demonstrate the usefulness – from the point of view of theoretical biology – of what Eden called the apparition of "dissymmetry" in cellular multiplication, Cohen planned to take into account the morphogenetic "density field", as it is known in embryology. This is the key to simulation's shift from combinatorial analysis to biological morphology. This concept of "field" had already been introduced in embryology in 1932 by Julian S. Huxley (1887–1975). In basing his own model on the physical concept of field, Huxley had hoped to generalize the notion of gradient (1915), which had originated with Charles Manning Child (1869–1954), by removing the bias toward a specific axial direction. This notion was then taken up by Waddington in order to explain the embryological phenomena related to this principle – which had been established in botany since 1868 – of growth towards the greatest available free space (Hofmeister's principle). If Cohen was to take this field into consideration using Eden's simulation technique, however, he could not make do with just a grid of square cells. He therefore came up with a *geometric* plane with a much more detailed spatial resolution that included the 36 points adjacent to the initial point of growth or of eventual branching, with each point separated by a 10° angle (since 36 × 10° = 360°). Each of these 36 directional points was affected by a "density field" calculated on the basis of the distances from each point to the other elements of the tree being constructed. Following Eden's extensive discretization, Cohen was thus obliged to make the space of biological morphogenesis geometric once again, since he could not otherwise see how a sufficiently realistic plant form could be designed if he retained such an over-generalized cellular approach. Instead, he reduced the mesh size of the grid, and above all retained the probabilistic formalization.

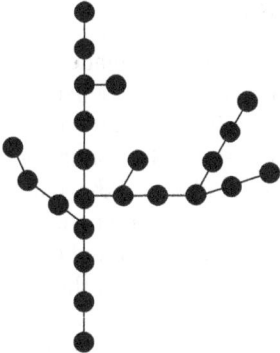

Figure 1.1 Tree-like shape generated using a Cohen simulation (1967) with directional bias favouring upwards growth.[3]

The length and angle of growth that took place from the starting point were determined by the density field. The branching rules were also affected by this density field, but were above all probabilistic. A test was carried out at each growth point by random number selection (pseudorandom), in order to determine whether or not the computer should add a branch in that direction. Finally – and it was here that Cohen could introduce his idea of a *programmed epigenesis* that was nonetheless sensitive to environmental events – he reused Eden's concept of variable branching based on the directions of the plane: this made it possible to simulate *heterogeneous* density fields arising either from the presence of other parts of the organism or from the potential pre-existence of a physical obstacle outside the organism. Using this programming flexibility, Cohen could also simulate areas that, on the contrary, facilitated growth. For Cohen, the resulting shape was conclusive, since it brought to mind experiments in mould growth on a heterogeneous nutrient substrate.

Cohen also chose to vary the growth and branching probabilities in accordance with the order of the branch in relation to the trunk. A branch attached to the trunk was order 1; a branch issuing from this branch was order 2, and so on. In this way, the hypothesis of hierarchically organized genesis could be tested, since the order represented a difference in biological status on the level of the general organization of growth. Since the diagrams that were generated by the plotters appeared to Cohen to resemble qualitatively realistic trees, leaf veins or even moulds, he considered that the results of the simulation were very conclusive. It had been able to rise to the two initial challenges of his theoretical epigenesis problem: 1) to prove the credibility of biological growth that is constrained by rules that have a fixed form, but whose parameters are at the same time sensitive to environment; and 2) to prove the credibility of the theoretical representation of this growth process as being hierarchically organized. This entire simulation process was published in the *Nature* journal in October 1967.

The epistemic functions of modular programming, simulation and visualization

As a result of the conditional branching made possible by computer programming, the model was able to incorporate a sensitivity to environment expressed as a feedback effect from the environment on the genetic parameters. These subroutines had the property of seeking the spatial optimum by self-adapting the rules of growth. Cohen then noted that these mathematical rules, which were constant but nonetheless had variable parameters, were extremely simple. The brevity of his FORTRAN program (only about $6 \cdot 10^4$ bits) merely confirms the feasibility of this type of scenario in nature: the small size of the program substantiates the initially counterintuitive idea that natural morphogenesis requires only a limited amount of elementary "information". Cohen even assessed the number of genes that would be necessary for the biological insertion of these $6 \cdot 10^4$ bits, and calculated that it would amount to 30 genes.[4] The briefness of the informational message was due to the modular nature of the programming, which avoided the repetition of computer instructions of the same type. Finally, it should be noted that, with this model, Cohen used the ability of MIT's computer at that time – the TX-2 – to *visualize* these calculations of point positions on a plotter so as to evaluate his project's success: he hoped to demonstrate visually that a hypothesis of growth that was both structured and epigenetic could apply to natural phenomena. For Cohen, as we can see, simulation by digital computer served essentially as a means of testing theoretical hypotheses. Similar ideas had already been expressed since the earliest days of digital computers, initially in work on nuclear physics, and later in biochemistry and physiology.

According to Cohen, two specific conditions had to come together in order to consider that a simulation could enable a biological theory to be rejected. In the first condition, the computer program itself must "incorporate" the hypotheses. To begin with, it materialized or embodied, so to speak, something that until that point had been merely spoken words. In this sense, it became closer to the empirical. Furthermore, the prevailing view of the times, which represented the genes as information units and the genome as simply a program, assisted in this identification. Next, the use of the plural term "hypotheses" must be noted: the program fleshed out a set of hypotheses and not just one isolated hypothesis. As we have seen, there was a variety of rules that replaced the uniqueness of one law. In this program, there were growth rules *and* branching rules: these two types of presumed rules were tested together, overlapping each other. For the second condition, it was necessary that a comparison could be made with natural shapes. In other words, the simulation could not have the power to discard hypotheses unless the computer could provide a means of comparison with the empirical: the simulation did not, of its own accord, reject a set of hypotheses; its ability to reject had to come, by transitivity, from the strong resemblance of its results to what is seen in nature. Thanks to the technical visualization system added to the TX-2 computer, however, the simulation could transmit its ability to discard morphogenetic theories to the computer – an ability that, until that point, had been the prerogative of actual or physically simulated experimentation.

It should be noted, nonetheless, that this comparison (and therefore this relationship of transitivity) was carried out here in a purely qualitative manner, or "by eye" we might say. It is no doubt for this reason that, unlike Eden, Cohen – who was well informed in the subject of biological substrates – was the first to use the term "simulation" to refer to this type of modelling and visualization of living forms by computer. Indeed, the title of his 1967 article is "Computer simulation of biological pattern generation processes". He owed his use of the term in this context to his acquaintanceship with the cybernetician von Foerster.[5] But according to Cohen, the occasionally qualitative nature of the comparison in no way diminished the unambiguous nature of the overall process of hypothesis testing, because he was alluding only to phenomena of shape that had already been identified as being typical, generic and easily recognizable globally and to the naked eye. In the case of a theoretical approach, the diagrams produced by the plotter evoked clearly enough – for those specialized in the field – real organisms that have been observed in real life. The diagrams thus did not prove the hypotheses, but nor did they reject them, and above all they might retain the hypotheses as being plausible despite having seemed scarcely likely beforehand.

At first, due to the inadequacy of the classic mathematical languages, the biology of shapes was unable to directly include the shape of living organisms in a single formal language so as to then attempt to reconstruct their evolution by means of an abstract model. For this reason, as we see in Cohen's work, the biology of shapes came up instead with local hypotheses – which for their part could be formalized by computer – about what generated these shapes, in order to make them construct what we see on an integrated scale: the overall shape. It was these atomic hypotheses (since they dealt with interacting elements) that benefited from processing and visualization by computer. The act of bringing experiment and theory closer was therefore not a form of abstraction but of solidification or of concretization. With simulation, rather than drawing the experiment towards the theoretical by stylizing and abstracting, the opposite movement (but which resulted in a similar convergence) was effected, bringing the theoretical closer to the experiment. Simulation brought the theoretical closer to the experiment by means of a concretization through a sub-symbolic use of the formal that translated – particularly in biology of shapes – into a requalification of the quantitative: a diagram that is locally interpreted but that can also be perceived and qualified globally by the naked eye.

Nevertheless, even though Cohen refined Eden's cellular approach, he was still left with a very imprecise consideration of morphogenesis. Even if he regeometrized, and if, in so doing, he could make the simulation converge towards a globally recognizable reality, he would still only be taking into account small parts of organs that he had cut away from each other according to what the formalism required for simple decisive recognition, but without any concern beforehand for their exact biological significance nor for their histological or physiological realism. This also explains why his stochastic, geometric and graphic simulations remained extremely speculative.

Another geometrizing approach, which was less speculative and more concerned with realism, would emerge, but in a more specifically botanical context, linked to two researchers; a Japanese physicist and an American botanist. These two researchers aimed to represent varieties of specific trees in a much more photorealistic manner. Let us look at the motivations behind their work, as well as its particular contributions.

The first geometric and realistic simulation of trees (Honda–Fisher, 1971–1977)

Although Cohen's proposal initially left agronomy modellers indifferent, one physicist from Kyoto University, Hisao Honda, adopted it explicitly in 1970. As a researcher in shape recognition, in the context of developing computer material specifically aimed at graphic applications, Honda was interested by the problem of how genes determine the genesis of living shapes, and in particular of branching shapes. His initial question was: "How is it possible that one can guess the species of tree on the basis of its form which is very variable and cannot be easily grasped in scientific terms?"[6] It so happened that Honda was technically able to deal with this issue as a physicist in Kyoto because, having been employed since 1969 at the University's Data Processing Centre, he had access to computer equipment that was particularly high-powered for the time: a FACOM 270–30 (Fujitsu) computer, with a top-quality graphic screen. This enabled him to represent the branches calculated in considerable detail.[7]

Honda's initial problem in this case was that of infra-linguistic recognition. In accordance with classic automated shape recognition, Honda recommended using tree "models"[8] that would be constructivist, step by step, with reiteration and with variable parameters. These models would be selected by a process of trial and error. The best of these generative models might be a model of the implicit process (non-verbal and not able to be verbalized, and therefore not referring to linguistic categories) of recognition of a tree by a person with a normal level of education. But this problem of shape-recognition that Honda set out to resolve in his physics department was, by his account, entirely parallel to the specific biological morphogenesis question of "how to describe economically the form or how to pull out the essence from miscellaneous information about the form".[9] Honda adopted a similar approach to that of Cohen because, after studying the first works of computer simulation of branching forms, he ultimately drew the same conclusion: simple generation rules can give rise to complex overall shapes that may vary considerably depending on the parameters of the rules. In 1971, therefore, Honda aimed to be even more specific than Cohen and demonstrate that an image of a "tree-like body" can be created on computer through the interplay between simple elementary rules. He did not choose a particularly detailed or complex model for his branching model, as otherwise the mental feasibility of shape recognition for such a model would be uncertain, as would the biochemical feasibility of information storage in the genotype. He chose to have access only to the branching angle and to the relative ratios of branch lengths after branching. But, unlike Cohen, he depicted the trees in three dimensions and not as flat structures. The FORTRAN computer program would project each of the tree's states on the screen in accordance with simple projection rules.

Honda's simulation was clearly based on an existing geometrical formalization, since it was aimed at the production of a realistic form on a graphic display screen. In an evolved language such as FORTRAN, management of the points or luminous dots on the computer screen is effected with reference to a metric orthogonal coordinate system. From a technical point of view, therefore, there was no need to translate the mathematical expression of the model in the program in order to obtain a visual representation. All Honda had to do was simply address the issue of transition from a metrics model in three dimensions to a two-dimensional representation, so that it could be seen on screen. Furthermore, as far as branching and branch growth were concerned, Honda – unlike Cohen – did not consider the effects of neighbouring trees. Using these deliberately rough hypotheses, since they made the elementary rules simpler, the computer seemed to be the best tool not only to deal with these simple and reiterated calculations, but also to visually represent their results. In so doing, Honda produced a clearly interdisciplinary work, since it was based on an analogy between the gene (with its assumed property of informational minimalism) and the minimal mental model of recognition of a branching shape.[10]

In Honda's work, the choice of the most credible models was made by an iterative process of trial and error. For him, if a credible shape of an existing tree species (ginkgo biloba, maple, birch, azalea, etc.)[11] could be roughly recognized on screen, then the parameters of the generating model, which were entered somewhat randomly into the computer at the start, should be retained or refined by further tests; otherwise, their numerical values should be rejected. In order to benefit from a more certain and informed ability in recognition, Honda was assisted by the morphometrician S. Oohata[12] from the Department of Forestry in Kyoto University's Faculty of Agriculture.

Somewhat later, in 1977, Honda further cemented his alliance with botany. As it happened, at the same time as he began focusing his work on the geometric simulation of tissue mechanics, the young botanist Jack B. Fisher – who at that time was employed at the Fairchild Tropical Botanic Garden in Miami – was working on the development of branching in a specific genus of tropical tree; *Terminalia*. As a botanist, however, Fisher worked in the field, and had access to a large number of measurements of real trees. In 1975, having learned of Honda's first article on the subject, Fisher soon recognized a possibility for fruitful collaboration: why not further refine Honda's theoretical architectural simulation to the extent of calibrating it on real measurements? To Fisher's mind, it would be very useful to have a precise description of tree crowns. Indeed, according to thesis that the American botanist H.S. Horn had just published in his 1971 book *The Adaptive Geometry of Trees*, the tree crown was to a large extent the site of the adaptation of trees, not only with respect to their environment but also insofar as other species of trees were concerned. Work towards calibrating a simulation on real trees would therefore make it possible for descriptive botany to offer more tools for precise prediction in the fields of forestry, silviculture and arboriculture. At the time, however, in order to distinguish between different plants according to their branching structures, botanists mainly employed a morphological and statistical approach that, in Fisher's view, seemed to have reached its limits of usefulness. In order to understand what led Fisher to opt for computer simulation and collaboration with Honda, rather than the statistical analysis that was predominantly used at the time, I must first briefly recount the limits that by then had in fact been reached, both in the domain of morphology and in that of the causal and physicalistic morphology of trees.

The limitations of morphometry and of thermodynamics of trees

One of the formalized approaches to the growth of branching shapes originated in 1945 with the study of fluvial geomorphology. Robert Elmer Horton (1875–1945), a Hydraulics Engineer at the Cornell University laboratories, had found that, statistically, for a specific fluvial "tree", there was a constant average ratio (approximately 3.5) between the number of branches located at a given branching order and the number of branches located at the subsequent order. He called this the "bifurcation or branching ratio". Likewise, he found that the "length ratio" was also fairly constant: the average length of a fluvial branch was 2.3 times greater than the average length of one of its branches.[13] Horton therefore broke down the ontogenesis of the tree, from a biometric perspective, in order to classify, count and measure its branches, branching order by branching order. As early as 1953, the geophysicist A.N. Strahler allocated numbers to the branching orders, starting from the tree's terminal branches. This numerical formalization could be applied to any type of tree, whatever its nature (whether botanical, fluvial, mathematical or decision trees), and made it possible to easily test whether "Horton's Law" applied equally to non-fluvial trees in the other empirical domains under consideration. Finally, between 1962 and 1971, the American physicist and geomorphologist, Luna B. Leopold, who was then Chief Hydrologist of the United States Geological Survey (USGS), undertook to find an explanation for Horton's phenomenological law in terms of the thermodynamics of open systems by using an approach based on structure optimization. In the same way as a fluvial network appears to optimize surface distribution according to its physical constraints, the botanical tree also appears to optimize its distribution of solar energy and the products of synthesis. Leopold's work thus consisted of trying to base this presumably interdisciplinary thermodynamic approach on a technique related to statistical physics.

At the beginning of the 1970s, however, a work appeared that would not only confirm Leopold's approach in one sense, but would at the same time clearly demonstrate the irrevocable limits of this use of Horton's Law in plant morphology. This was a work by three biologists and physiologists from the Department of Medicine at Birmingham's Queen Elizabeth Medical Centre in the United Kingdom. Leopold had tested Horton's Law on a single botanical tree. In 1973, however, while revisiting Leopold's problem, which involved testing for the existence of "branching ratio" and "length ratio" (which in this case would be the "diameter ratio" of the branches) in botanical trees, Barker, Cumming and Horsfield demonstrated that Horton's Law was also statistically applicable to apple and birch trees.[14] They based their study on actual and exhaustive field data, however, and not on partially estimated data as Leopold had done: they chose two real trees – an apple and a birch – and counted, measured and ordered (according to branching order) each of the branches of the two trees.[15] Their results confirmed the possibility of transferring Horton's Law to plant morphology.

As a result of their use of the computer as a data storage and processing tool, however, which enabled them to effectively take into account the *entirety of the information measured in field in its fundamental variability*, what distinguished their study from Leopold's was the reflection elicited by their resulting histograms. They observed that there was considerable variation in some of the geometrical parameters for certain branching orders. This was the case for the apple tree in particular. There was such great variability of size at times, that the lengths and diameters of the branches were not characteristic of any specific order: it was therefore possible to introduce an error in branching order if the measurements were not carried out on every single branch. In order to obtain correct ratios, it was necessary to save the entire metrical information – the information, in other words, that allowed the spatial structure of the real tree to be reconstructed, branch by branch. So it did not seem likely that the original objective of this type of formalism – i.e., the reduction of information for the purpose of theoretical usage – would be achieved for botanical trees. In 1973, it seemed that the thermodynamic and statistical approach for botanical trees would not allow morphometric data to be correctly reduced and condensed without damage, particularly insofar as biologists and botanists were concerned.

The first geometric simulation of an actual tree: *Terminalia*

Jack B. Fisher was firmly convinced that, if the aim was to be able to take into account the tree-crown geometry that was so crucial for theoretical study of the adaptive strategy of the tree in the field as well as for studying its energy, then the possibility of completely and geometrically reproducing the shape and structures of this crown would have to be retained, otherwise false global values would result. This meant that the crown characteristics could not be quantitatively and validly summed up using such morphometric techniques, or at least not as far as botany was concerned. Nonetheless, like Honda with his initial problem of shape recognition, Fisher in turn was also seeking a minimal means of reproducing the design of the crown without losing too much of its complexity, because he was in fact looking for the key parameters that were linked directly to the genome and that determined the optimal form of tree crown for the purpose of capturing solar energy. Thus, contrary to Honda's expectations, it was not exactly a geneticist who answered his hopes and took up the gauntlet he had thrown to biology in 1971, but rather a biologist working on an already much more integrated scale: the morphological and developmental scale. Fisher and Honda began to collaborate in 1976, albeit initially from different locations. Honda was obliged to make his generative and geometric model somewhat more complex so as to be able to directly use the averages of the measurements that Fisher made in the field. Fisher, for his part, felt that they should begin with *Terminalia*, since its crown presented distinctly differentiated leaf stages, giving it a fairly simple pagoda-like structure. In his 1971 book, H.S. Horn had already pointed out the advantage of studying such trees for anyone wishing to quantitatively and easily take account of the role of the tree crown in the tree's adaptive strategy. But, according to Fisher, this simplicity was still not sufficient unless a computer-simulation approach was added to it. Indeed, despite the relative simplicity, it was no longer possible to conduct the calculations for parameter optimization analytically in cases where it was not possible to analytically summarize the characteristics of the structure resulting from the various geometrical parameters chosen at the outset. Therefore *the entire tree-crown must be simulated* by computer.

In 1977 Fisher and Honda published the first realistic geometric simulation of a *Terminalia* of less than five years of age. In order to do so, Honda designed an Assembly-language program on an Olivetti P652 computer equipped with a memory extension and a curve plotter.[16] The results were issued in the form of graphs on paper. In the "mathematical model",[17] which formed the heart of the program, they used a set of deterministic geometric rules whose parameters were first set to the averages of Fisher's field measurements. At the beginning, therefore, the program drew an average tree, so to speak, whose bearing clearly brought to mind, intuitively, the form of the trees found in Japanese parks or in the Fairchild Garden in Miami.

In 1978 the authors decided to extend their collaboration in order to test what use might be made of simulation in the case of adaptive geometry. The problem here, of course, was finding an optimum. Since the tree crown could not be summed up analytically but could only be replicated by simulation, it was therefore necessary to carry out a great number of realistic simulations based on the various possible combinations of parameters around their actual averages, and to measure afterwards each time – i.e., on each of the plotted structures – the resulting efficiency at capturing solar energy of each of the tree crowns obtained. This research was carried out by trial and error, more or less empirically: Fisher and Honda ran a large number of tree-crown simulations, each time varying (within limits that were considered to be realistic and close to the range of averages measured in the field) the branching angle of branches inside the leaf clusters, using the computer each time to calculate the surface area exposed to the sun. It should be noted in passing that the simulation thus replaced one calculation by two distinct operations: a replication followed by a measurement. In this way, they found an optimal branching angle for exposure to sunlight, given the morphological constraints of the species under consideration.[18] As they had hoped, the resulting value was found to be very close to the actual average value. Honda and Fisher demonstrated that this was also the case for the optimal "length ratio".[19] The actual tree was therefore reasonably optimal from an energy point of view in relation to its geometric growth rules. This rather remarkable result was the subject of a publication in the February 1978 *Science* journal.[20] In view of these results, in fact, it could be considered that geometric simulation, with its few simple parameters, had managed to grasp an underlying structural variability that was the aim of selection for a functional optimum in *Terminalia*. Nonetheless, Fisher was aware that it would be too simplistic to link the determinism of the tree-crown architectural parameters to just the optimization of leaf surface. The issue of optimization must be much more complex overall. At most, all they had demonstrated thereby was the obviously important role of this particular optimization in the first years of life in *Terminalia*, especially since the geometric model was no longer valid for older trees.[21]

In 1979 an article appeared that had a great impact in plant morphology and would contribute to reviving botanists' awareness of the differential ageing of the various plant parts and of the mutability of their growth rules over their lifespan. This article was entitled "The plant as a metapopulation", by the Irish plant demographer James White (Department of Botany, University College in Dublin), and appeared in the *Annual Review of Ecological Systems*. James White was driven by the issues he had encountered as a population biologist. Since he was used to seeing populations in the world of plants, where the difference between population and individual often remained problematic, he did not claim to have invented the suggestion of perceiving the plant as a colony of more or less autonomous individuals. In fact, this concept may be seen in many works from the 18th century onwards. In this synoptic and seminal article, however, White first outlined a fairly complete conceptual history of this population-based concept of individual plants and then forcefully demonstrated the obvious weight it should hold for botanists and plant morphologists in the future by highlighting the areas of convergence that were then becoming clear. This point of view, which was already proving fruitful in population biology, but was now explicitly and clearly set out by White, contributed at that time to further highlighting the variability of genetic determinisms within a plant individual during its ontogenesis, i.e., during its lifespan. In calling for a determined merger between plant demography and plant morphology, White stressed the malleability of genetic determinations during morphogenesis.[22]

Thus, if a tree was also a population that had a considerable genetic plasticity in its organs and during its life, then Honda's genetics- and informational-based argument (along with Cohen's argument, for that matter) that formed the bedrock of their common approach using a simulation based primarily on a formal stationary model would lose much of its pertinence. It seemed that the search for a minimal theoretical model that would determine morphogenesis with complex and counterintuitive results – since morphogenesis always occurred under the pretext of a genetic determination that was assumed to be immutable during the plant's lifespan and its sequence of cellular differentiations – was no more valid than other theoretical approximations along the lines of "Horton's Law".

Fisher recognized that this was a further limitation of computer simulation: in fact, it would be necessary to make the subjacent mathematical model even more complex by incorporating the historicity of the genetic determinisms, as well as including the effects of environment. But the problem was precisely that increasing the complexity was not desirable for the use he wished to make of simulation. The theoretical uses of the simulation would thus be lost: it could no longer be used to produce simple results regarding the nature of the very few parameters he hoped would emerge as being the only decisive ones in the development of the plant's adaptive strategy, for example. Simulation could no longer serve to designate, or even to quantify, those few key parameters. If the underlying model was made more complex, which of course was always possible in principle if not in practice, then it would no longer be possible to use the simulation as a theoretical argument because it would no longer give immediate help in reaching a better physiological and functional understanding of the structure. Yet this was precisely what Fisher favoured. And it is here that one can clearly see the implicit epistemic role that Fisher attributed to tree-growth computer simulations: they were a simple extension of the theoretical expression of biological concepts, and a means of rigorously testing their pertinence in specific cases. The replication of the tree was not sought in its own right. It was not even really considered to be a substitute for experimentation, but rather as a substitute for the random procedure of optimization of a complex mathematical law, like a calculation. For Fisher, the realistic depiction of the tree was thus subordinate to his aim of an understanding that would ultimately be able to leave behind its technological and computational medium in order to swell the body of biological and botanical knowledge that was already expressible in natural language.

Thus, from 1979 onwards, Fisher began noticeably to gradually cease active collaboration with computer plant simulators, considering that simulation could not offer him any increase in biological comprehension.[23] By his own account, simulation might at most still look the part by resorting to increasingly complex models. But these models would remain overly simplistic for too long from a biological point of view. For that matter, computers required sufficient computational power to deal with this, and it did not really appear to him that this was the case. After a period of keen hope, therefore, Fisher entered a phase of unswerving lack of conviction regarding the pertinence of computer simulation for theoretical biology.

A recap of geometric simulation

The geometric simulation approach was well and truly the first to converge genuinely and quantitatively towards the botany and biology of higher plants. It helped to make possible some of the deductive calculations involved in testing certain general hypotheses specific to biology and theoretical developmental biology. Because of the intertwining and feedback between multiple causal chains during the morphogenesis of living organisms, this deductive follow-up is not in fact easily understandable in human thought, whether formalized or not: there is no analytical mathematical model of it, and the results may therefore be counterintuitive. The only solution is to use the computer to simulate the maze of relationships between parties and between the parties and the organism as a whole, so as to be able to follow it step by step.

But when we know that the plant is a "metapopulation", when we see clearly that the causal chains themselves are evolving (e.g., ageing of the meristems[24]), then a further level of complexity is added. It was permissible, therefore, to doubt the benefits that simulation itself could offer to theoretical botany. If the local rules changed over time, would an approximation obtained by computer recursivity still be valid? Was there a rule for the changing of rules? And how could it be found?

Notes

1 Cohen (D.), "Computer simulation of biological pattern generation processes", *Nature*, 1967, Vol. 216, 21 October, p. 246.
2 Ibid.
3 Author's own adapted and simplified diagram.
4 Ibid., p. 248.
5 Private correspondence of 20 December 2006.
6 Honda (H.), "Description of the form of trees by the parameters of the tree-like body: effects of the branching angle and the branch length on the shape of the tree-like body", *Journal of Theoretical Biology*, 1971, Vol. 31, p. 331.
7 Ibid., p. 335.
8 Ibid., p. 332.
9 Ibid., p. 332.
10 Ibid., p. 337.
11 Ibid., p. 335.
12 In 1971 the biologists S. Oohata and T. Shidei published an article on the branching structure of trees, but they carried out only a statistical (biometric) analysis of the size of the branches in relation to their order of branching.
13 Leopold (L.B.), "Trees and streams: the efficiency of branching patterns", *Journal of Theoretical Biology*, 1971, Vol. 31, pp. 341–342.
14 Barker (S.B.), Cumming (G.), Horsfield (K.), "Quantitative morphometry of the branching structures of trees", *Journal of Theoretical Biology*, 1973, Vol. 40, p. 33.
15 Ibid., p. 34.
16 Fisher (J.B.), Honda (H.), "Computer simulation of branching pattern and geometry in *Terminalia (Combretaceae)*, a tropical tree", *Botanical Gazette*, 1977, Vol. 138, No. 4, p. 377.
17 Fisher and Honda's own term, ibid., p. 378.
18 See Fisher (J.B.), Honda (H.), "Branch geometry and effective leaf area: a study of *Terminalia*-branching pattern, 1. theoretical trees", *American Journal of Botany*, 1979, Vol. 66, No. 6, pp. 633–644, and Fisher (J.B.), Honda (H.), "Branch geometry and effective leaf area: a study of *Terminalia*-branching pattern, 2. survey of real trees", ibid., pp. 645–655.
19 Honda (H.), Fisher (J.B.), "Ratio of tree branch lengths: the equitable distribution of leaf clusters on branches", *Proceedings of the National Academy of Sciences of the USA: Botany*, 1979, Vol. 76, No. 8, August, p. 3875.
20 Honda (H.), Fisher (J.B.), "Tree branch angle: maximizing effective leaf area", *Science*, 1978, Vol. 199, No. 4331, 24 February, pp. 888–889.
21 Fisher (J.B.), Honda (H.), 1979, art. cit., p. 639.
22 White (J.), "The plant as a metapopulation", *Annual Review of Ecology and Systematics*, 1979, Vol. 10, pp. 133–134.
23 Private correspondence with Fisher dated 28 October 2003.
24 See Glossary.

2 The logical model and algorithmic simulation of algae

By the end of the 1960s, however, it seemed that an alternative might be feasible by refining the formal representation of the rules of growth. Inspired by the intellectual movement dating back to the 1930s that promoted the logical axiomatization of any formal theory in science, a logical and mathematical approach was developed that was initially intended to help explain morphogenesis. When this approach was put into practice, however, it also ended up being in the form of computer simulations, but this time as algorithmic simulations, through formal automata controlled by computer programs. As a result, the biology of shapes would lay claim to another use of computers, in particular through the work of Aristid Lindenmayer (1925–1989). Despite some similarities with Dan Cohen's approach, the type of formalization required by Lindenmayer's approach was based on a completely different field of theoretical biology and above all on a completely different interpretation of what mathematization, formalization and theorization meant for biology. Although the embryological and developmental issues that Lindenmayer dealt with seemed close to those of Cohen or Fisher, he did not use the computer in the same way as they had: instead, he sought to avoid any "simulation" in the sense of Cohen's stochastic realism or Honda and Fisher's geometric realism. He proposed a form of modelling by automata (later known as L-systems) that would nonetheless increase the possibilities for using computer simulation, and give rise, as we will see, to a decisive controversy regarding "natural formalisms". How, then, should we view this third type of simulation used in biology (after Cohen's essentially probabilistic simulation and Honda and Fisher's essentially geometric simulation) in plant morphogenesis?

A botanist won over by logical positivism: the "theory of lifecycles" by A. Lindenmayer (1963–1965)

During the 1963 university year, the Hungarian botanist Aristid Lindenmayer, who was at that time attached to Queens College of the New York University, spent some time in London with the British embryologist and philosopher of science Joseph Henry Woodger (1894–1981). He had been invited there on a university fellowship funded by the National Science Foundation. Woodger was 69 years old when Lindenmayer visited him, and was by then retired. For many years his presentations had been essentially philosophical in nature. Nevertheless, Lindenmayer had heard of his work and hoped to establish a regular intellectual and interpersonal relationship with him, even though

DOI: 10.4324/9781315159904-3

his theoretical biology work had been more or less abandoned since the late 1930s. In 1932 Woodger had been one of the co-founders of the Theoretical Biology Club in London, which had been established around the ideas of d'Arcy Thompson and Alfred Whitehead on the continuity between biology and physics, and which was where Woodger often met the biochemist Joseph Needham and the embryologist Conrad Waddington. It was in this context that his anti-mechanist or anti-reductionist stance emerged, along with what he called his "methodological vitalism".[1] According to Woodger, just as quantum mechanics had led physics to relinquish stolid realism with regard to the ultimate constituents of matter, so, in the same way, theoretical biology, without falling into pure phenomenalism or fictionalism, ought to base itself on different "realms" of realities by establishing explanatory and axiomatized theories that could demonstrate how to pass from one realm to the other, without any expectation of ever finding a realm of fundamental realities. Furthermore, Woodger considered that biological language was too often used for extra-scientific purposes, including emotional aims, for example. A formally purified artificial language that could be subjected to calculation would neutralize these detrimental subjective tendencies. This was the main reason he advanced in *The Axiomatic Method in Biology*[2] to justify his adhesion to the logical reconstruction schemes of logicians and philosophers such as Rudolf Carnap. In this work in particular, adopting the extensional and logicist approach of Russell and Whitehead's *Principia Mathematica*, Woodger started from the possibility of considering any organism or organ as being constituted of temporal and spatial "slices". From these "atoms" of the living being, he developed a general set of axioms for biology in which he demonstrated, in essence, that, contrary to the affirmations of the vitalists, for example, it was possible to formalize a one-to-many relation, or a relation of rising complexity and cellular differentiation, in an organism during ontogenesis. In other words, by using a formalism that drew a distinction between types at the same time as their axiomatized regulated relation (following Russell's theory of types model), the increase in degree of complexity of a whole – i.e., its development – could be represented, according to Woodger, without resorting to a vital principle. In this respect, it should be noted that Woodger did not acquire any models from the nascent mathematical theory of the time, although he had in the meantime met Alfred Tarski and had shared a number of his reflections during a period when Tarski was developing his model-based alternative to the axiomatic conception of logics and mathematics. Consequently, Woodger's work, inspired by an axiomatic approach that was no longer new, initially met with little acclaim and the approach was embraced by philosophers rather than by embryologists, even theoretical embryologists.

In the early 1960s Lindenmayer's main biological focus was on what he called – in accordance with an idea that had come to him during discussions with an American botanist, Ralph O. Erickson (1914–2006), who was then a professor at the University of Pennsylvania – the "life cycle theory".[3] At that time, Erickson was a renowned botanist, working alongside plant physiologist David Rockwell Goddard (1908–1985) in particular. Erickson had been among those who, in the face of the classical statistical approach to issues of growth (based essentially on average phenomena and focused primarily on the already high level of the organ), had begun to advocate a theoretical approach to morphogenesis, addressing individual cell behaviour. This type of approach may, in fact, have appeared more promising, since it would then be possible to follow the generation and filiation of cells – their "lifecycle" – even on a cellular level, but without having to standardize their locally differentiated behaviours. But the necessary mathematical and formal tools to do this were lacking. Erickson, for his part, did not have an educational background that might predispose him to fundamentally changing his approach to morphogenesis in this way. For this reason, starting in 1965, he instead adopted the formalism of partial differential equations in order to try to account for these locally differentiated morphogenetic behaviours. It was Lindenmayer, therefore, who in 1964 first suggested an alternative formalism by resurrecting Woodger's earlier logical approach. Lindenmayer's epistemological training, together with his unique mathematical skills, would contribute considerably to the birth of a new formalism that, right from the start, would prove to be particularly well adapted to computer simulation, even though its original aim was theoretical.

Lindenmayer considered that the history of any cell or any cellular nucleus, and thus – by extension – of any living being by virtue of its organogenesis, could only be described by a particular combination of three successive processes starting from a primitive cell or a nucleus: mitosis, meiosis and gametic fusion. Lindenmayer was particularly knowledgeable about these issues because he was originally a specialist in fungi and algae. Algae have one of the greatest numbers of different types of cellular reproduction, and Lindenmayer aimed to use these to form a veritable deductive theory. Woodger's logical approach appealed to Lindenmayer more than any other, because it allowed the formal language to disregard the cytological and biochemical complications that occur during each of the three processes, unlike the earlier approaches to growth of Goddard or Erickson, for example. Only what Lindenmayer called the "cardinal events"[4] of the lifecycles would be focused on. Without a doubt, he was also incited to adopt this formalist view of biological theory because he also shared some of the opinions of Woodger and of logical empiricism on the notion of scientific theory. For Lindenmayer and Woodger alike, a theory is a particular type of formal language that is used for recording observable facts and for syntactic deduction, within that language, of the symbolic representation of those same facts or of other, so-called, foreseeable facts based on primitive notions and axiomatic postulates or rules. For that matter, Lindenmayer borrowed directly from Carnap not just the distinction between syntax and semantic (which Carnap had in fact inherited from Tarski, even though Lindenmayer did not cite him), but also the idea that, because of this distinction, science – speaking sensibly (i.e., semantically) of the world of phenomena – must establish rules of correspondence between observations and symbols of formalism, or, in other words, between the semantics or meaning of the concepts symbolized in the language and the theoretical concepts involved in the formal propositions of the theory. Lindenmayer called these "semantic rules".[5]

In contrast to Woodger's generation, and therefore seemingly in agreement with Suppes'[6] view, Lindenmayer appeared to have learned from the mathematical theory of models. Yet he was not always consistent in this claim since, because he was influenced by the theoretical aspects of biology and morphogenesis, he also insisted on the theoretical nature of what he called his "mathematical models" in the sense that, as we shall see, he always hoped to draw theorems from them.

In 1964, with his three primitive notions (mitosis, meiosis and gametic fusion) acquired, all that remained for Lindenmayer to do was to try to produce the postulates or axioms required for formally deducing the different types of lifecycles observed in nature by biologists. Having defined his symbols and invoked several elementary theorems from *Principia Mathematica* and Carnap's symbolic logic, which he would in fact use, Lindenmayer inserted into his axioms certain biological rules that were well established among biologists, and in particular among botanists, and that dealt with the issues of succession and of combining the three elementary processes of generation. Certain combinations were excluded. Furthermore, according to Lindenmayer, when the biological material involved in each of these processes was disregarded, it was possible to represent each of these processes as a multilateral formal relation,[7] i.e., as a relation of one to many or of many to one: the mitotic relation was thus a relation of one to two, the meiotic relation was a relation of one to four, and gametic fusion a relation of two to one.[8] Lindenmayer's theory therefore clearly aimed to deal with what we have seen to be one of the major difficulties in formalizing organic life and its development (and which had been previously recognized by Waddington and Rashevsky): the fact that relations are rarely binary. At this point, it can already be seen that, from the point of view of cellular generation or even of the organic development of metazoans, the theory of combination of lifecycles appeared to be able to produce a more adequate formal representation.

Unusable set of axioms and used set of axioms

Lindenmayer revived Woodger's axiomatic principle of relations between cells[9] after decisively modifying it. First of all, he relinquished the concept of organic "slice" altogether, since Woodger saw this as being both a spatial and *at the same time* a temporal demarcation. This view, which was in theory too general, led to confusion since it did not determine which of the two dimensions – temporal or spatial – would offer the true or best *biological meaning* in the theory. In Woodger's axiomatic system, Lindenmayer noted, "the zygote from which an animal develops and the gametes to which it gives rise, as well as the cells in between, are all parts of the same whole organism".[10] Using Woodger's rather surrealistic definition for the organic "part of . . . " relation, it becomes very difficult to later insert in a useful manner – or in other words, in a semantically feasible way – the necessary restrictions that could give a real biological significance to certain types of partitions of a real organism. Woodger had formalized the biological object from a very high and generalized viewpoint (in a way that in fact remained closely bound to a particular outdated technical perspective) with the aim, so he thought, of letting nothing in these formal symbols escape that might one day be liable to biological interpretation or, in other words, to what he considered observation (epistemological descriptivism). In so doing, he had advocated a style of carving up biological reality that was practically useless as it stood. This was the important lesson that Lindenmayer taught Woodger. In reality, once the organic parts considered in Lindenmayer's formal system were ordered next to each other in accordance with a specific combination of the three elementary generation processes, they became temporally organized *at the same time*: there was no need to add a formalization of time that would make proving the theorems unfeasible. Lindenmayer was thus able to prove the theorems for the "lifecyles", because he had shaped his axioms beforehand in order to achieve this end. He even calibrated his axiomatic system directly on the type of organic parts (and therefore on the type of partition) that interested him. The payoff of this preliminary choice can be seen in the resulting non-generalizable nature, as far as biology was concerned, of the new axiomatic principle he proposed. As a consequence, Woodger's theoretical bird's-eye view perspective was abandoned.

Finally, in summing up on this first (1964) of Lindenmayer's theories, it should be noted that its publication in a chapter of a collective philosophy of science work in honour of Woodger did not do much for its subsequent recognition, especially among theoretical biologists, very few of whom ever heard of it.[11] The theory was not taken up or extended in its original form, even by Lindenmayer. For that matter, he never subsequently referred to it again, except in his seminal article of 1968,[12] as he considered that his first real contributions to theoretical biology only appeared from 1968 onwards. So what actually happened that year that was so special?

From logical theory to automata theory (1966–1967)

To find out what happened we must look back to 1966–1967. During this period Lindenmayer was still a Researcher and Professor of Biology with the Queens College Department of Biology at New York University. He was still pursuing his work with the assistance of a grant from the National Institutes of Health, which in turn formed part of the US Public Health Service. As a side note, this means that the National Institutes of Health was a pioneer in this type of research, since they had previously also supported the work of Murray Eden and Dan Cohen. Lindenmayer's focus was still on algae and their development, both in the sense of "lifecyles" as well as from a developmental biology point of view, i.e., in a morphogenetic sense, in particular. In this respect, Lindenmayer, in stark contrast to Woodger, headed a biology laboratory where practical experimentation rightfully held a well-deserved place.[13] In the three years since his "theory of lifecyles" had appeared, Lindenmayer had learned of the existence of the theory of automata, in particular from John Richard Gregg, who at that time was Professor of Zoology and Theoretical Biology at Duke University and a long-time associate of Woodger.[14] Gregg gave Lindenmayer the idea that he could perhaps use this new formalism in the representation of organic development. Furthermore, in 1967, NASA requested that Lindenmayer oversee urgent research that was due to be carried out at the Institute of Theoretical Biology that had been established at

Fort Collins, Colorado, for pre- and post-graduate students. The main aim of this Institute was to promote research – in keeping with the methods of cybernetics and of the automata theory – in exobiology (the study of extra-terrestrial life) and to thereby encourage students to engage in the field at an early stage.[15] In this context, NASA chose the British psychiatrist and cybernetics pioneer William Ross Ashby (1903–1972), who was then Director of the Department of Electrical Engineering at the University of Illinois, to head the group. Ashby was tasked with providing an introductory seminar on cybernetics. Lindenmayer, for his part, was invited to bring his research assistants, and therefore attended with two of his most motivated students from the preceding semesters: Andrew Schauer and Jerome C. Wakefield.[16] While attending, he was requested, and conscientiously began, to dissect all the recent literature on automata modelling of life, spurred on, in particular, by the hopes raised by Ashby's books, such as the 1952 *Design for a Brain*, or *Introduction to Cybernetics*, which had been published in 1956. Together with Schauer and Wakefield, therefore, Lindenmayer read through all the available literature of the time. It should be recalled that, in 1966, a posthumous edition of Von Neumann's work on cellular automata had also just been published under the supervision of Arthur W. Burks (1915–2008), which contributed to making this research on automata truly accessible to the widest public. Lindenmayer quickly read through these works and thus spotted the theoretical foundation underpinning automata. From 1965 to 1967 he and his students at Queens College, of New York University, had worked on cybernetic models over many months, but without ever opting directly for integrally discretized representation. But it was a seminal book, also published in 1966, that was primarily responsible for introducing Lindenmayer to such representation: *Cybernetics and Development*. This monograph was written by a British zoologist and psychologist, Michael J. Apter, who had followed a very eventful and interdisciplinary career path. He had presented his doctoral thesis on the links between cybernetics and developmental biology at Bristol University. He had first worked in Lewis Wolpert's Zoology Laboratory in the University of London's King's College, at a time when Wolpert was studying both experimentally and on a cellular level the morphogenesis of the sea-urchin embryo.[17] Subsequently, Apter joined the Department of Psychology at the University of Bristol, where the Automation Engineer, Frank Honywill George, introduced him to cybernetics. It was at that point that Apter changed direction for his thesis and was instead steered by Professor J.L. Kennedy of Princeton University's Department of Psychology towards developmental biology models. His aim was to draw lessons for psychology from the new cybernetic models that were then being proposed – albeit somewhat timidly – in developmental biology. At that time, according to Apter, the cybernetic approach made it possible to go beyond not just physicalism (including Rashevsky's first works in biophysical mathematics) but also the mathematicisms of Woodger or Gerd Sommerhoff.[18] According to Apter, these earlier approaches had failed in their attempts to analytically resolve the problem of the irreducible nature of biological phenomena. To his mind, only cybernetics offered a formal conceptualization (negative feedback) that would provide a general non-vitalistic explanation of the things that affect not just living beings, but also teleonomic machines. The success of Jacques Monod and François Jacob was proof of this. Despite their claims, Rashevsky, Woodger and Sommerhoff had stopped short at description, whereas cybernetics promised an explanation. To explain this point, Apter based himself explicitly on the distinction that logical empiricists, influenced by Tarski, made between syntax and semantics: cybernetics proposed a general and formal principle on a syntactic level; furthermore, it was a source of particular models, which were thus, in turn, valid on a semantic level. For Apter, as for Waddington, an organism undergoing the process of genesis was comparable with a complex network of rules that could be represented formally by the interaction of self-reproducing automata. The novelty in Apter's approach was his decision to discretize. In particular, he revived Waddington's concept of "Pattern Formation" in order to try to theoretically model the creation of structure on a unidimensional cellular tissue with a geometry similar to that of the hydra. To do so, he represented the biological cells with Turing-type automata: the cells, which were arranged end to end, changed state and the information that they passed to their neighbouring cells depended on their present state and on the information they received from the right and from the left. All the cells bore the same rules: to this extent, they all had a common "genotype", as Apter called it. But their state changed: this was their variable "phenotype".[19] The cells did not give birth to other cells, and the tissue did not grow: as in Turing's morphogenesis model, it was assumed to pre-exist. There was, however, communication between cells, and Apter demonstrated, by carrying out manual calculations, that there was stabilization of a certain heterogeneity in relation to the distribution of the cells' different internal states. His model was therefore similar to Turing's, although he committed himself to a discretized approach right from the start. Whereas Turing, in his 1952 work, *discretized a* chemical and mathematical *model*, Apter *discretized the substrate from the outset*.

Thus, in 1967, Lindenmayer saw clear signs of a converging trend among mathematicians (Ulam and Eden), and among certain cybernetic biologists or biophysicists (Apter), with increasing use of the digital computer to represent the growth of organic forms in a mathematical but discretized manner. Since he considered the theory of automata to be closely linked to mathematical logic, and especially to symbolic logic, he felt that the representations produced by automata disconcertingly matched his own way of conceiving formal representations of organic development. Moreover, since he aimed to model *organic development* – a phenomenon in which substance expands in space – Apter's cybernetic model on the *formation of structures* was not exactly suited to his needs. The existing automata models, on the other hand, offered the possibility of *extending the formal structure*.

At the same time, Lindenmayer also wished to examine the dynamic behaviour of cells both during their formation by division and in their interactions within the multicellular organism as a whole. He considered it essential to try to take the *interactions due to the contiguity of the cells in multicellular beings* into account. In this matter, he followed the American plant physiologist John Gordon Torrey (1921–1993), who considered that the link between molecular biology and developmental phenomena on the level of the organism could only be understood through an emphasis on "intercellularity" in research. Torrey was a specialist in cellular differentiation and in root growth under the effect of plant hormones. According to Torrey, the cells of multicellular organisms continuously exchange energy, pressures or metabolites, and, in return, these exchanges decisively determine the cells' physiological and metabolic behaviour. It was therefore unrealistic to believe (as theoretical biologists such as Rashevsky once did) that an understanding of the development of superior organisms could be achieved by simply and directly combining the behaviour of several unicellular beings.

Since it would be necessary to take spatial proximities into account, Lindenmayer considered it necessary – contrary to the requirements of his earlier "lifecycles" theory – to consider the organism's morphology or, at the very least, its topology, in order to enable the formalism to integrate these relationships of proximity between cells. It is here that Lindenmayer's work differs from that of Robert Rosen (1934–1998). Whereas Rosen – a student and colleague of Rashevsky – always started from the whole organism or even from a single cell and then used the theory of automata to roughly depict the metabolism and the cellular repair that was assumed to have occurred in certain localized areas of the organism (nucleus, cytoplasm, etc.), Lindenmayer proposed on the contrary to start his formalized representation on the level of distinct cells so as to make them effectively generate from and interact with each other. Rosen, in accordance with the spirit of Rashevsky's theoretical biology – and basing himself on the mathematical theory of categories – concentrated on the logic of metabolism and repair because what fundamentally interested him was trying to produce a mathematical representation of what was assumed to be the distinctive feature or essence of life. Lindenmayer's formalism, on the other hand, unlike Rosen's, was influenced by the earlier problem of "life cycles" and was instead conceived to take cellular division into account. Nonetheless, it was also in explicit continuity with Rosen's work that Lindenmayer attempted to propose a formalism that, as far as possible, allowed the a priori demonstration of theorems without having to use a computer. Or else, if the computer must be used, it would be primarily as a deductive machine, i.e., as a support in conceptualizing the consequences of the axioms, rather than to graphically represent them. I propose to call this *algorithmic simulation*.[20] According to Lindenmayer, we should not expect that the computer will present emerging physical properties that cannot be formulated or foreseen in the formal system, but rather that the computer will remain simply an infallible and powerful logical calculator (a deductive enumerator). It was for this reason that, unlike Eden or Cohen, Lindenmayer did not initially use simulated randomness in his algorithmic models. The computer gives us support in the job of deduction – a job that should by rights (if not in fact) remain the work of the human mind because of its unremittingly linguistic and logical nature: it is therefore necessary to harness the generating power of the machine and not aim to fill it beforehand with geometric depictions of physical and biological phenomena, but rather with logical representations. In this instance, the computer is a *machine for conceiving, but not a machine for imagining*. Thus, since Lindenmayer's theory was, for that matter, indebted to Woodger's axiomatism and logicism despite its occasionally semanti-cist claims,[21] it initially owed more to the computer theory itself, i.e., the theory of automata, than to the actual use of the computer as a simulator.

The "developmental model" and the rules of rewriting (1968)

In order to produce a formal model of the growth of such organisms, Lindenmayer therefore used the mathematical theory of sequential machines for the case of filamentous organisms (multicellular organisms with a filament or branching structure). His "sequential machine", *representing* a single cell, was a quintuple comprising:

• the transition function giving rise to the next state of the cell;
• the transition function giving rise to the next output of the cell;
• state variables;
• input variables;
• output variables.

The cell inputs corresponded to the outputs of neighbouring cells: this is the formalization of the intercellularity. But what was new and decisive compared with Ulam or Apter's earlier formalizations, for example, was that Lindenmayer allowed that the next state of a cell could be a *splitting*, i.e., a *division*, in biological terms, of that same cell. By so doing, he sought to formalize new growth that could initiate in any part of an already formed organism. Thus it was not the environment that made the cells spontaneously appear, as in Ulam's algorithmic model, but the existing cells that chose whether to multiply or not. In this way, the organism could grow. In order to express this model, Lindenmayer also broke with the *Principia Mathematica* representations that, because they were primarily linguistic and non-intuitive, were purely linear in nature; instead, he adopted a more graphical mathematical representation. He even confirmed this evolution by also adopting a way of representing these functions with what he called, in the manner of automation engineers, "tran-sition diagrams".[22] In these diagrams, the cell states were represented by the vertices of a graph and the various directed edges of the graph represented the possible transitions between states as a function of the cell inputs. Finally, it would not be possible to use this simple automata formalism of states and transition functions unless a process to determine the structure of time (i.e., the instants when these functions must be synchronously applied) was also given. This is why such a formalism, besides requiring a realistic discretization of the parts of the organism – the cells[23] – also requires a *discretization of time*: as with the simulations of Eden and Cohen, at each step of time each cell would apply its two transition functions.

Next, Lindenmayer had to modify the usual formalism of the transition functions so as to be able to represent the filament growth by division of a mother cell into two daughter cells. Here is his transition matrix for the cell states:[24]

		Cell input	
		0	1
Cell state	0	0	1
	1	11	0

Figure 2.1 Transition matrix for cell division (after Lindenmayer 1968).[25]

The content of the output squares in the table indicate the value (0 or 1) of the cell's next state as a function of its present input and state. In the lower middle square we see two states as there are now two cells, for which Lindenmayer specified the states at the moment of their creation. In this case, each of the daughter cells started with state 1. The first result obtained was not a photorealistic image, but a table with one column containing 1s and 0s, in which the daughter cells took the place of the mother cell. This gave a linear model in the form of a binary sequence. Next, in order to represent branching into several filaments, the precise spot where the branch will form must be indicated on the formal filament, along with the spot where the end of the description of the branch cells will occur. To do this, Lindenmayer decided to use parentheses. In this way, for a given cell state, the transition function of this state could be added to the state of this same cell *plus* the state of a new cell, in the same way as previously when only cellular division was possible, but this time adding the *internal state of the new cell between parentheses*; this signified that the cell would begin a side branch on the filament. The cell between parentheses itself might then divide or branch, but everything it produced would remain between parentheses, in order to denote that it was, and remained, a branch. The parentheses themselves could then be placed between further parentheses since there could, of course, be several orders of branching. The formalism of the parentheses thus had the effect of retaining the formalism's linearity, even when the filament had branched. The problem with this formalism, however, was – as Lindenmayer admitted – that the relative position of the branches, i.e., their mutual arrangement or phyllotaxis, was not taken into account. The spatiality of actual branching was not considered.

Lindenmayer then carried out an initial comparison between the model and a specific living species; a red algae called *Callithamnion roseum*. He discovered the important fact that the task of calibrating the model was very different from that in the case of a differential model. Indeed, Lindenmayer found he could base his calibration on just a few of the rules followed locally by the main parts of the developmental structure. And these rules were, in fact, precisely those explained in biological and histological works in (admittedly very technical) terms that could be easily translated into simple graphical terms and simulated by the sequential machine model. Thus, in an article published in 1971, Lindenmayer was able to assert that "[s]uch developmental descriptions [produced through finite mathematical methods] appear to be closer to our intuitive understanding of an organism".[26] This formalism therefore seemed closer to direct intuition of observed natural phenomena and their verbalization. It should be noted that, in the context of this application to the representation of red algae, Lindenmayer admitted to having actually carried out a "simulation" himself for the first time; in other words, when he sought – as he did here – to calibrate his general modelling infrastructure on this actual real-life algae, to replicate it, so to speak (even though he does not use such a term in his work), so as to create a discrete model with precise rules suitable for his purposes. For the red algae under consideration, for example, there are four known botanical and histological rules:

1   the main filament should have at its base one to three cells that have no branches;
2   then, each subsequent cell in the filament should, on the contrary, give rise to a branch and there should be only one such branch;
3   at each step, the four – or at the very least three – cells below the tip of the main filament must not bear any branches; and finally,
4   each branch order greater than or equal to 1 should in turn repeat the same three rules to itself as those applied previously to order zero, or in other words to the main filament.[27]

Lindenmayer therefore had no problem expressing these rules directly by transition rules for his automata-cells: the formal translation was immediate, since the morphological description already appeared as a genetic and logical account of how the shape was established.

For the first time, Lindenmayer represented the results of the calculation of the first few steps of this first model of red algae in a spatialized – and in that respect non-linear – form. In other words, he abandoned the formalism of parentheses and instead presented the results in the form of a hand-drawn branching

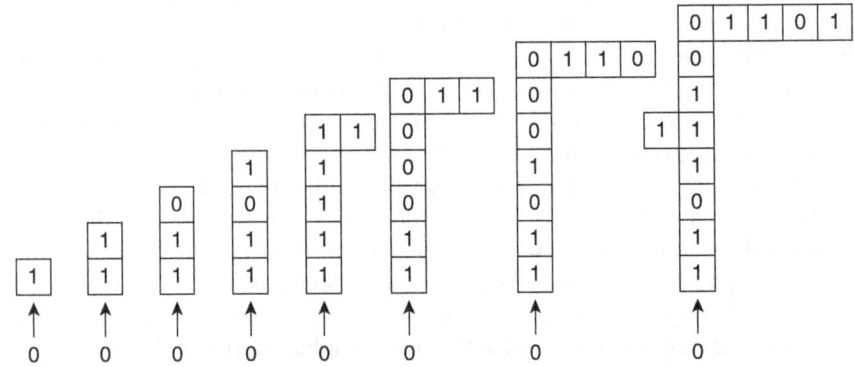

Figure 2.2 Principle of Lindemayer's logical growth and branching model.[28]

diagram, where each cell was now realistically represented by a square displaying the number of the corresponding state of the cell. The branches were placed alternately (and arbitrarily) to the left and then to the right of the main filament, since the formalism of the parentheses, as we recall, did not decisively determine this issue. The shift to a diagrammatic drawing in this regard would require perfecting, although Lindenmayer did not consider it essential at that time. He applied this extra rule of alternation simply in order to make the linear formalism graphically representable in a more realistic manner. Even though it had not been his initial aim, Lindenmayer in fact ultimately considered that the possibility of moving to a drawn format would be a very valuable asset for biology.[29] This possibility of translating the formalism into a drawing – a possibility that was, itself, linked to the fact that the formalism was more or less on the same wavelength as the biologists' intuition – was finally understood to be essential.

Moreover, Lindenmayer was aware that a purely formal consideration of such rewriting rules at the level of the entire cell could not be used directly to distinguish certain hypotheses in the field of chemical embryology regarding the mechanisms of induction or differentiation, for example. In particular, in the case of red algae, a model with information flows (chemical or otherwise) gave the same results as a model without such flows. But Lindenmayer added that, from this point of view, experimental embryology could not yet help the modeller to come to a decision. The theoretical progress that he felt he was offering would have to await similar experimental progress. Nonetheless, this type of theoretical work, that involved not just proving the simple *possibility* of a calibration and convergence towards experimentation, but also highlighting the fact that a formalism could give rise to *competing parameter settings*, was – according to Lindenmayer – indispensable in order to encourage and direct empirical investigation with the aim of preparing it appropriately for its role as arbitrator.

The dispute with Brian Carey Goodwin regarding "natural" formalisms

Unlike the earlier plant-shape simulation proposals (such as Cohen's), Lindenmayer's work immediately elicited a significant response. There is therefore plenty of material for analysis regarding the reception of "L-system" formalism, as it soon became known. This reception would contribute to the trend towards convergence between (theoretical) simulation and plant biology, but yet without being the actual "seed" around which – at a later date and in a completely different research context – the plant simulation method would precipitate and crystallize. What were the reasons behind this? It is necessary first to examine the way Lindenmayer's work was received in order to identify these reasons. On the one hand, Lindenmayer was quickly gripped by the fascinating technical complexity of the formal domain that his axioms had revealed. In a certain sense, his approach was initially, and for a long time, swallowed up primarily by technicians of formal languages and theoretical computer science. On the other hand, and rather naturally, certain theoretical biologists of the Waddington school of thought were among the first to consider themselves affected and to react to this unprecedented proposal to mathematize plant morphogenesis. But the confrontation with the British organicists was, in fact, the subject of a lively dispute that was also very instructive from the point of view of each party's technical and epistemological choices. Without dwelling on the theoretical works in linguistics and language theory that were prompted by L-systems, which would take us too far from our field of interest, I will concentrate on the debate regarding "natural" formalisms that took place in the form of a series of alternating publications between Lindenmayer and the theoretical biologist Brian Carey Goodwin.

Brian C. Goodwin (1931–2009) was born in Canada and initially studied biology at McGill University in Montreal. He then went on to study mathematics, under a Rhodes Scholarship, at Oxford. In 1961 he received his PhD in embryology from Edinburgh University, under the supervision of Waddington. His thesis set out an outline of a general theory of development and evolution. In essence, he retained his teacher's idea that morphogenesis could not be entirely explained by genetic reductionism. Like Waddington, Goodwin advocated the use of mathematical models that could demonstrate how living shapes are more constrained than the purely historicist and probabilistic scenario advanced by evolution theorists might lead one to think. Also like Waddington, he therefore sought not to refute but to amend Darwinism through a view of embryology that was mathematical and organicist, i.e., neither reductionist nor purely holistic. After a period as researcher at McGill and MIT, Goodwin became Lecturer in Biology at the University of Sussex in 1965.[30]

In the meantime, Goodwin had published a book in 1963 that met with a certain degree of success – *Temporal Organization in Cells* – in which, taking inspiration from Jacob and Monod's 1961 model of regulation, he proposed a mathematical model of simultaneous expression of several genes located in a network. This "network of genes" model described the simultaneous variations in concentrations of several proteins induced by gene expression in a cell. A number of multiplying terms, which revealed the concentrations of other proteins, were involved in the expression of each protein's variation in concentration (a first derivative with respect to time); this then made it possible for the other genes to regulate this speed. Goodwin thus ended up with a system of coupled differential equations that were rather difficult to solve. Thus, at the beginning of the 1960s, Goodwin was already fascinated by control models, and to his mind autocatalytic sets that could be expressed by coupled differential equations should be considered "as natural models of functional integration".[31]

Understandably, however, when Goodwin read the work that Lindenmayer published in 1968 in the *Journal of Theoretical Biology*, he reacted rather negatively. In 1970 he published an article on biological stability, in which he implicitly targeted Lindenmayer by openly criticizing the choice of a formalism of automata in developmental biology.[32] His argument boiled down to perceiving what was presented as a formalization by automata as not being a *real formalization* – i.e., an effective formal representation (or "natural" in the sense that Goodwin gave to his qualification of differential "natural models"), or even a close approximation – but rather a *simple analogy* comparing the gene with the computer that, in certain critical cases, failed to include certain essential biological phenomena. The formalism was therefore, for Goodwin, not truly a formalism: it was simply a "formal analogy". In his own words:

> The implication of the computer analogy is that the cell computes its own state, looks at the DNA program for further instructions, and then changes state accordingly.
>
> This is not in fact what a cell does, although a formal analogy can be made between the biochemical behaviour of a cell and the operation of an automaton following a program. It may seem elementary to insist that all the operations of the automaton must at some point be interpreted in biochemical and physiological terms, when discussing such a process as epigenesis, but I have been somewhat dismayed at the amount of confusion that has arisen because of a failure of those using the computer analogy to illustrate the operation of algorithmic instructions at the biochemical level.[33]

Thus, for Goodwin, this reliance on the automata formalism derived from a consideration of purely superficial and rough similarities between the inter-automata relationships and the relationships that take place between biochemical substances in the epigenetic processes. In this instance, "formal" therefore meant "superficial". According to Goodwin, however, the proof that this was merely a rough analogy could be inferred from the confusion in which adherents of automata modelling found themselves when required to "illustrate the functioning of the algorithmic instructions on a biochemical level". In turn, this confusion was not shown directly, but could be seen in the confusion that reigned around that illustration. Goodwin therefore pointed out that there may be several different biochemical illustrations for the same set of automata model parameters. In other words, for Goodwin, this "formalism" did not have a sufficiently detailed and unequivocal grasp of biochemical phenomena to be truly free of ambiguities. Whence the confusion. Whereas Lindenmayer, as we have seen, considered this formalism to be almost *too detailed* to be calibrated unequivocally using the available experimental data, Goodwin, on the contrary, considered it *not detailed enough* to have unambiguous biological significance. In this sense, according to Goodwin, the formalism demonstrated the classic flaw of analogies: they distort reality at the unexplained (or "neutral", we might say, in the words of American epistemologist, Mary Hesse[34]) edges of the analogy relationship and can thus lead to arbitrary decisions, which is a source of confusion, about what part of the actual correlate is represented, or not represented, by the analogy at its uninterpretable edges.

Lindenmayer's response was to demonstrate that, in fact, it is up to the user of this type of model to remove any ambiguity that might slip into the interpretation, since "the concept of finite automaton is general enough to give it the desired interpretation in biochemical and cell-physiological terms".[35] On the contrary, far from being an obstacle, it was the formalism's generality that allowed interpretative ambiguity to be removed, according to Lindenmayer. He asserted that each of the *controls* (or regulators) "can be expressed as either turning on and off the genes, or as affecting the activities of the enzymes and thus the appearance of their products".[36] He therefore sought to interpret these points of least agreement in terms of automata. Here, in essence, is what he asks his detractor to recognize and assume:

- Let *C* represent the set of metabolites and components of a cell, except for the active proteins and nucleic acids (which Lindenmayer calls "informational macromolecules"),
- at any given moment, we can assign the *state* of a cell by a particular *combination* of the elements of *C*,
- each "gene-enzyme" pair can be associated with a "*transformation rule*",[37]
- and finally, each cell can be associated with a *combination of transformation rules* depending on whether the corresponding genes are active or not at the moment under consideration.

Under these conditions, the *total internal state* of a cell will consist both of the *combination of its components* and the *combination of its rules of transformation*. The *input* of a cell will consist of the set of components, elements of *C*, that have newly entered the cell during that specific (discretized) time interval. The *output* will consist of the set of those components that have left the cell. We can thus see that, in conformity with the concepts of developmental biology, the *next state* can only be determined by the present state, the inputs and outputs. Thus, it is possible in fact to construct discrete transition functions that will determine the next combination of components in each cell in the same way as the next combination of transformation rules. Lindenmayer concluded this passage as follows:

> It can be seen that both the next-state and the output functions can in principle be constructed for a given population of cells, thus the cells can be validly represented by finite automata. I cannot agree, therefore, with Goodwin's objections. What he calls the 'DNA program' is the set of production rules we introduced, and according to our short discussion above it can be said in a completely natural way that the next state of a cell is computed by using these production rules.[38]

We are faced here with an obvious and indicative confrontation between two different interpretations of what ought to be a natural mathematical model. For Goodwin, the naturality lay on the side of differential models because the actual correlate of the formalism remained ambiguous; for Lindenmayer, it was to be found on the side of automata because they had the ability to precisely and without ambiguity translate existing theoretical concepts; for him, this was the naturality of comprehension, it was the proximity of a sign compared with its meaning, whereas Goodwin sought the naturality of a reference and thus emphasized the closeness between the formal sign and the actual referent. In a sense, we can say that the naturality of Goodwin's model is ontological, whereas Lindenmayer's, in accordance with the logicism he inherited from Woodger, is gnoseological and symbolic.

In Goodwin's criticism we can thus already see one of the areas of resistance to the use of automata in both modelling and simulation in developmental biology. Unsurprisingly, one of the sources of this resistance was to be found in the Waddingtonian and organicist school of thought. After Turing's model, and even before the revival of these attitudes by certain theoretical biology proponents of the 1980s, the resistance was already well established. In addition to the fact that knowledge of the automata theory spread rather slowly in biological fields, it must also be admitted that Lindenmayer's approach did not immediately take root in biology: quite the contrary. Nevertheless, there were some exceptions, such as the theoretical work on tissue growth by the French researchers Hermann and Jacqueline Lück, of the Laboratory of Analytic Botany and Plant Structuralism (*Laboratoire de Botanique Analytique et de Structuralisme Végétale*), Saint-Jérôme Faculty, Marseilles, who discarded the statistical biometry approach of the early 1970s in order to adopt and spatialize L-systems. After this necessarily brief depiction of the origins of automaton modelling of branching shapes, I will now carry out a rapid review.

Recap: the computer as automata model and deductive machine

What can be said, in the end, about simulation and the epistemic use of the computer in this sort of automaton modelling? Realistic simulation is often invoked in work in this field, but only by way of theoretical and logical argument: in reality, it is only rarely put into action and carried out to completion. *It is more the possibility of simulating plant logics or plant "language" than the actual simulation itself (in the sense of displaying the computation results) that is emphasized.* For that matter, the models are often validated before the simulation is even carried out. Because its structure made it likely to readily produce an algorithmic simulation, logical modelling by automata (or rules of production) was, in its early stages, interesting *in its own right* because it made it possible at times to express directly in the form of theorems (i.e., in a theoretical and abridged form) a certain number of a priori results that could be calculated by hand and verified by experiment, without necessarily requiring computer simulation of the model. Thus, at the beginning, logical modelling was seen above all as a new formal means of expressing a theory and deducing its predictions, even though this delegated formal technique was backed up by roughly realistic representations that could already be described as (algorithmic) simulations in this sense.

It should not be forgotten that, around the same time, a different theoretical use of the computer and of the possibilities of simulation displayed, on the contrary, a certain concern for a more faithful representation of reality, in this instance of trees. This was the approach of H. Honda and J.B. Fisher. We have seen, however, that their approach was soon faced with the problem of flexibility in the local rules of morphogenesis. Furthermore, Lindenmayer's automata technique, even if it could in fact have done so, did not initially appear capable of easily formalizing this flexibility in practice, since its initial advocates were primarily seeking to produce theorems.

Lindenmayer's solution could not be adopted unaltered by botanists specializing in trees and in higher plants in general (i.e., plants with vegetative organs). In the end, it was instead in an agronomic context that this issue of the flexibility and evolutivity of local morphogenesis rules in higher plants was finally addressed and partially resolved. For this to happen, however, it was necessary for researchers in simulation to learn to use all available means and avoid all preconceived bias towards any preferential formalism. It was also necessary that such researchers should not feel constrained by any theoretical preferences (premature generic explanations) or overly restrictive epistemological choices (i.e., simulation reduced to an assigned theoretical or symbolic cognition). Only then would it be possible to carry out combined simulation, without hindrance. Might it be possible to effectively integrate these different views of simulation so as to be able to draw even closer to botanical reality? But, in order for this to happen, there would have to be a need, a demand, so that it would not also end up being purely speculative.

In order to understand precisely how such diversity and such inclusive and operational convergence finally came about, in contrast to these other more theoretical conceptions and uses of simulation, my historical analysis will focus more closely on the emergence and development of *architectural simulation* – a type of simulation that could also be called "universal", since its formal structure made it valid for any type of plant. This focus can be justified by at least two reasons: on the one hand, while research into mathematical phyllotaxis could certainly boast of relying on certain observations, only architectural simulation could ultimately formalize the plant as a whole; on the other hand, it was the only approach that made it possible to quantitatively calibrate the complete formalization of the plant and thus make it capable of being used in the field. *It was thus architectural simulation alone that truly brought the history of plant-shape modelling into a new era: that of convergence with practical work and problems in the field.* At the same time, I will also give an account, as necessary, of the alternative proposals, both speculative and pragmatic, that have continued to mark these past five decades. But I will give these less weight than previously, precisely because the epistemological proposals that emerged from them were not innovative in view of what happened during the same period in the field of architectural simulation. In particular, these alternative proposals cannot explain the relative ascendancy that the school of architectural simulation has now acquired. It will provide an opportunity to confirm on the evidence the variety of different epistemic uses of formal models, as well as how such uses have changed since the emergence of the computer, especially with regard to their particular links with actual experimentation and with statistical experimentation (experimentation *by means of* a statistical model), as well as with biological concepts – some of which have emerged refined and rectified in the process.

Notes

1   Roll-Hansen (N.), "E.S. Russell and J.H. Woodger: the failure of two twentieth-century opponents of mechanistic biology", *Journal of the History of Biology*, 1984, Vol. 17, No. 3, p. 420.

2   Woodger (J.H.), *The Axiomatic Method in Biology*, Cambridge, Cambridge University Press, 1937.
3   Lindenmayer (A.), "Life cycles as hierarchical relations", in J. R. Gregg and F.T.C. Harris (Eds), *Form and Strategy in Science: Studies dedicated to Joseph Henry Woodger on the Occasion of his Seventieth Birthday*, Dordrecht, D. Reidel Publishing Company, 1964, p. 417.
4   Ibid., p. 465.
5   Ibid., p. 416.
6   Cf. introduction.
7   Although Lindenmayer did not use a generic term for this in English, the author uses the term "*relations formelles multilatérales*" (multilateral formal relations) in French to denote all the types of relations between individuals that are not binary, i.e., only bilateral.
8   Ibid., p. 434.
9   See Woodger (J.H.), op. cit., section 4, pp. 85–86.
10 Lindenmayer (A.), art. cit., p. 466.
11 One notable exception is Stahl (W.R.), "The role of models in theoretical biology", *Progress in Theoretical Biology*, 1967, Vol. 1, No. 1, pp. 165–218.
12 Lindenmayer (A.), "Mathematical models for cellular interactions in development. I. Filaments with one-sided inputs", *Journal of Theoretical Biology*, 1968, Vol. 18, pp. 286 and 299.
13 Jerome C. Wakefield confirmed this information in a private correspondence (e-mail) dated 30 September 2003.
14 See Gregg (J.R.), Harris (F.T.C.) (Eds), op. cit., p. 4. In 1953, in keeping with Woodger, John R. Gregg had published a work applying first-order logic to taxonomy and classification systems. Gregg also made several contributions in the field of logics.
15 From private correspondence with Jerome C. Wakefield, dated 29 September 2003.
16 In 1967 Jerome C. Wakefield already held initial diplomas in philosophy, mathematics and psychology. He later qualified as a psychologist and became known for his work in artificial intelligence, carried out in collaboration with Hubert Dreyfus whose critical views he shared.
17 See Apter (M.J.), *Cybernetics and Development*, Oxford, Pergamon Press, 1966, pp. x (Preface), 39 and 53. In the early 1960s Wolpert resurrected the "developmental mechanics" of German embryologist Wilhelm Roux (1850–1924). Roux had demonstrated that simple physical changes effected on sea-urchin embryos could have physical consequences for the organism.
18 In 1950 Gerd Sommerhoff, an anatomist at Trinity College, proposed what he called Analytical Biology. For Sommerhoff, this involved using analytical functions to take account of the great number of biological terms describing purposeful behaviour: psychology could thus be conceived as an extension of biology. Like Apter, he later belonged to the group of cyberneticians (most of whom were British) called ARTORGA: The Artificial Organisms Research Group. This group existed from March 1958 to December 1974. The scientific committee included, among others, Ross Ashby and Heinz von Foerster. The group brought together psychologists, physiologists and specialists in social and behavioural science.
19 Apter (M.J.), op.cit., p. 135.
20 Of course, any computer simulation uses algorithms and may be called algorithmic in this sense. But this term is intended to denote a type of use of the computer based on a small set of homogeneous rules (rule-based); this set of rules forms an explicit algorithm which, as such, has the particularity of directly representing certain behaviours of the living system in question.
21 As we have seen, this ambivalence was already present in Apter's work.
22 Lindenmayer (A.), art. cit., p. 282.

23 This discretization was already present in Woodger. But it was in Lindenmayer's theory of life cycles that the discretization became based on the direct, and in this sense realistic, identification between a symbol and a biological individual or a very distinct part of that individual from the point of view of biological knowledge (the cell). The formalism thus became mildly symbolic under these conditions.

24 Lindenmayer (A.), art. cit., p. 285.

25 Author's own adapted and modified presentation.

26 Lindenmayer (A.), "Developmental systems without cellular interactions, their languages and grammars", *Journal of Theoretical Biology*, 1971, Vol. 30, p. 455.

27 This is a summary of the description in Lindenmayer (A.), "Mathematical models for cellular interactions in development. II. Simple and branching filaments with two-sided inputs", *Journal of Theoretical Biology*, 1968, Vol. 18, p. 309.

28 Author's own adapted and simplified diagram.

29 Lindenmayer (A.), 1971, art. cit., p. 475.

30 In 1983 he became Professor of Biology with the Open University in Britain. He left this post in 1996 in order to join the teaching body of the Schumacher College, a centre for the study of theoretical ecology based on the concept of *deep ecology*, on James Lovelock's Gaia Hypothesis and on essentially holistic views on ecology.

31 Stuart Kauffman was citing his friend Goodwin's words from memory in Kauffman (S.), *At Home in the Universe. The Search for the Laws of Self-Organization and Complexity*, Oxford, Oxford University Press, 1995, p. 274: "those autocatalytic sets are absolutely natural models of functional integration".

32 For this information, see Lindenmayer (A.), "Cellular automata, formal languages and developmental systems", in P. Suppes, L. Henkin, A. Joja, G.R.C. Moisil (Eds), *Logic, Methodology and Philosophy of Science IV, Proceedings of the 4th International Congress for Logic, Methodology and Philosophy of Science, Bucharest, 1971*, Amsterdam, North Holland Publishing Company, 1973, p. 679.

33 Extract from Goodwin (B.C.), "Biological stability" in C.H. Waddington (ed.) *Towards a Theoretical Biology*, Vol. 3, Edinburgh University Press, 1970, pp. 3–4; quoted by Lindenmayer (A.), 1973, art. cit., p. 679. In this quotation Goodwin was specifically criticizing a similar stance originating from the theoretical chemist Christopher Longuet-Higgins (1923–2004).

34 Mary B. Hesse was emphasizing the heuristic role of mechanical models in this way. Cf. Hesse (M.B.), *Models and Analogies in Science*, Notre Dame, University of Notre Dame Press, 1966; 2nd printing: 1970, p. 8. In this sense, Goodwin only believed here in the negative impact of the undetermined part of the analogy.

35 Lindenmayer (A.), 1973, art. cit., p. 679.

36 Ibid., p. 680.

37 Ibid., p. 680.

38 Ibid., p. 681.

3 The limitations of biometric models and the transition to simulation in agronomy

This third method of simulating plant morphogenesis by computer (after the geometric method and the logical and algorithmic method) was first conceived and developed in the early 1970s by a French agricultural engineer, Philippe de Reffye. Unlike the earlier methods, it evolved from a purely pragmatic need. There were a number of more or less fortuitous, yet linked, causes behind its emergence. Contrary to the cases of the earlier authors, these reasons were no longer solely personal, speculative, rhetorical or aesthetic; instead, they were based on a combination of technical, institutional and political motivations. Since this work was no longer merely speculative, it involved the efforts of a number of researchers, although de Reffye was undeniably the main instigator for much of it. After I have reviewed de Reffye's principal pioneering suggestions, therefore, I will assess the work of his laboratory's followers. We will then be able to see how the fields of agronomy and forestry felt the need to go beyond the mathematical model stage and enter the stage of computer simulation. We will see, above all, that the strength of simulation was based on three things: 1) the decision to proceed by what I have called fragmented modelling; 2) the possibility of creating an integrative computer simulation – based on these fragments once they have been reconstituted by computer – that can be calibrated in the field; and 3) the possibility of taking the spatial and temporal heterogeneity of plant-part growth rules into consideration and thus bypassing the limitations of traditional biometric models.

The institutional and technical context of the IFCC (1966–1971)

Philippe de Reffye was born in 1947. In 1971, having graduated as an engineer from ENSAT (*Ecole Nationale Supérieure d'Agronomie*[1]) in Toulouse and holding a DEA[2] in genetics and plant improvement from the Orsay Faculty of Sciences, he set out on a five-year national-service aid work assignment at the Bingerville[3] research station of the *Institut Français du Café, du Cacao et autres plantes stimulantes*[4] (IFCC). The IFCC stemmed from the earlier "Coffee, Cocoa, Tea" service at ORSTOM.[5] From the outset, the service was designed to foster the rational and systematic deployment of programmes for improvement in yield, in particular using the new genetic methods for cultivated crops such as coffee, cocoa or tea, in what was known (until 1958) as the French Union (comprising French Equatorial Africa, French West Africa, Madagascar and New Caledonia). Improving the

DOI: 10.4324/9781315159904-4

yield of coffee plants had been long neglected. At most, it had been left to the initiative of experimental gardens: occasional improvements had been achieved but were often ineffective, since there was no point in identifying a more productive strain if it then turned out to be more susceptible to diseases or drought. It therefore became increasingly clear that it would be necessary to overcome or control a number of factors at the same time, in a rigorous manner over a period of many years. But according to an initial report by René Coste, who had been the first director of "Coffee, Cocoa, Tea", the members of this new service were all agreed on the fact that this type of undertaking was no longer accessible to untargeted private initiative. At the end of the 1950s, therefore, the French agronomists more than ever promoted the idea that, as far as crops from the French tropical territories were concerned, the complexity of their field of study required an organization that was dedicated solely to plant improvement, with a strong emphasis on genetics and biometrics. If increased coffee production was desired, then very strict centralized follow-up, management and evaluation programmes would have to be set up: the time for haphazard plant-variety selection had passed. Thus what, in 1900, had initially been a "test garden" at Bingerville, and then from 1929 became an "experimental station" devoted to coffee-plant and palm-tree breeding, similar to the famous experimental station at Rothamsted where R.A. Fisher had worked, now became a research centre specialized in the genetic improvement of coffee and cocoa plants in particular.

Thus, thanks to this new set-up as a research centre, when de Reffye arrived at Bingerville he was able to immediately avail himself of the new "Arabusta" hybrid that had been developed and stabilized between 1961 and 1971 by J. Capot (who was initially Agricultural Engineer and later Head of the Genetics Division at IFCC) and his team. In order to introduce the genetic inheritance from the Arabica into the Robusta (which was more adapted to the Côte d'Ivoire climate) while retaining the advantages of both varieties, the IFCC geneticists had hybridized the two. At that time, however, Arabusta's yield remained consistently inferior to that of Robusta. The IFCC coffee-plant improvement specialists had therefore only half-fulfilled their mission; it was still necessary to establish a policy for selecting the tetraploid parents involved in creating the Arabusta plants so as to ultimately obtain coffee plants that were both adapted and more productive. Such a policy did not seem easy to define a priori, however, since there was *no simple morphological characteristic* that could serve as a unique marker for future yield. "It was not sufficient to just weigh the harvests in order to compare productivity",[6] de Reffye wrote later. It was therefore necessary to consider a set of morphological characteristics and see how their various combinations were correlated with yield.

As a matter of course, the initial approach to the problem was essentially biometrical, since numerous factors and their co-evolution had to be taken into account. Indeed, it was from this perspective that de Reffye first chose to address the issue of coffee-plant productivity, making full use both of his knowledge in statistics and of the available literature at IFCC.

Transferring a little bit of econometrics to biometrics: a problem of optimization (1974)

Nonetheless, de Reffye immediately chose to view this problem in terms of seeking a constrained optimization. The classic methods of biometry, as his colleagues were already aware, were not always satisfactory in practice in Côte d'Ivoire: such methods could not explain certain surprising empirical results. For a given coffee-plant clone, for instance, the foliage levels might give contradictory results from one year to the next. Implicitly including these levels in the statistical model therefore gave unsatisfactory results. According to de Reffye, it was first necessary to be able to find what he called the "optimum in plant improvement".[7] Descriptive models did not seem suitable when, as was the case here, the aim was not just to study the variability of a population by means of principal component data analysis,[8] but also to find an optimum that might not yet even have been obtained from that data. What was wanted was a model that made it possible to choose the optimal crop-row direction for yield. For the same reason, the multiple regression models – which de Reffye called "predictive models"[9] – while of course allowing extrapolations based on the data, did not make it possible to clearly determine a hypothetical optimum. De Reffye therefore used a constrained optimization linear programming technique inspired by econometrics models. The "limited variability" and the "correlations" between the morphological traits of the coffee plant would represent the "constraints" in the linear programming of coffee-plant production, in an econometric sense, while the yield – expressed as a "linear model"[10] function of the characteristics – would represent the "criteria" to be maximized. This method transfer from operational research was possible partly because all the morphological characteristics that de Reffye used were of a numerical nature (it was a principal component analysis, not a factor analysis), and in part because he considered it admissible to hypothesize that the plant's yield could be expressed as a linear function of its characteristics. He used five such characteristics, divided into two groups. The first group included the characteristics of the coffee-plant leaf: 1) leaf shape (length by width); 2) leaf size (the square root of its length multiplied by its width); 3) leaf density (the leaf mass divided by its surface area). The second group included two branch characteristics: 4) branch thickness; and 5) the number of nodes on the branch. De Reffye's method of considering coffee-plant "yield" and productivity was thus identical to that of economists when dealing with problems of optimum production in businesses; de Reffye's aim was likewise that of finding a production optimum. In his first model, the plant was thus viewed as being analogous to a factory. Since from an agronomic perspective, and more precisely from the point of view of plant improvement, it was also necessary to decide on a policy (for the selection of plant variety), it seemed likely that the analysis and decision-making model that was first used for artefacts (the factory or human production) could also be transferred to a "natural"[11] object, such as a plant.

The first numerical result demonstrated that traits 1, 2, 3 and 5 were closely correlated to each other, whereas trait 4 (branch thickness) varied independently. Increase in thickness thus appeared to be unrelated to architectural growth (represented in this case by leaf growth and number of internodes[12]). Next, de Reffye examined the newly expressed data as a function of the principal components, and then expressed the yields in terms of this new frame of reference. But these figures also showed that there was no correlation between the yield regression vector and the principal components: *the yield was therefore not expressed simply.* This meant that the principal components that had been established numerically on the basis of the aforementioned five main traits were still too imprecise and not sufficiently informative to make it possible to predict yield. Furthermore, coffee-plant clones or families present significant intrinsic variability when it comes to yield, as has been well documented. Thus, while the traits were obviously responsible for yield, they were themselves difficult for the breed-selector to control. Nevertheless, when the yield regression vector was expressed as a function of the initial variables (the five numerical characteristics of the coffee plant), the most significant component was seen to be the fifth: *the number of nodes per branch.* The leaf traits were thus of little importance insofar as yield was concerned, and the branch thickness, for its part, played only a moderately important role, even though thick branches are generally more productive than thin ones. As far as coffee-cherry production (and thus coffee-bean yield) was concerned, it became clear that functional and mathematically simple allometric[13] relationships could be ruled out.

Furthermore, the importance of breakage in over-ramified branches was recognized a posteriori from the results. Such branches broke under their own weight; their daughter offshoots were therefore starved of nutrients due to reduced sap supply and this impacted negatively on their fruit production. It thus became apparent that non-linear phenomena could interfere in coffee production and limited the relevance of an approach using linear models, even when these were multiple-factor models. As a consequence of these results – and this was decisive for what followed – de Reffye recommended that, in future, the plant "habit" should also be taken into account, particularly since a certain number of "good producers had a tendency to bend, and were not adapted to commercial crop activity".[14] This first work ultimately demonstrated that it was unrealistic to seek an optimum directly by means of linear statistical modelling, even if it remained possible to find a regression model by calculation: the breed-selector would obtain neither precise knowledge nor any new grasp of the phenomena in this way.

What is more, the existence of a frequent phenomenon of fruiting failure required a better understanding of the factors controlling fertility, in particular those factors relating to the flower and its pollination. This was precisely de Reffye's aim in the second work on modelling he produced at IFCC. He therefore abandoned his first work on the search for a global optimum, which he subsequently almost never cited in his later works. His first approach had thus been more or less a dead end, but he learned a lesson from it that he would never forget: even though a solution may seem elegant, or even merely practical and effective, it is unrealistic to aspire to immediately and uniformly formalize complex living phenomena such as the morphogenesis of higher plants.

The first application of plant simulation in agronomics (1974–1975)

The second 1974 article alone contained the core of de Reffye's post-graduate[15] thesis, which he defended in 1975. In this article, the operational research-type linear programming approach was abandoned. De Reffye also discarded his yield-based global approach and decided to sort the successive phenomena leading to fruit production by order of occurrence. This time, he studied the events of fructification in detail by identifying the precise criteria of good fruit production in coffee plants. He realized that he needed to understand and master fructification by the *most detailed possible reconstruction* of the process leading to fruit bearing, although in doing so he would have to sacrifice the simplicity of his mathematical solution. He therefore concentrated first on just the fertility of the ovules in the final stages of fructification, i.e., on the development of the beans inside the coffee cherry. All he had to do was trace back just a few of the final processes that took place before the coffee-bean harvest, and find observable traits that would allow him to forecast the harvest just before it took place. The aim of de Reffye's first PhD thesis was thus to demonstrate that it was possible, first of all, to find such observable traits and, second, to match these observable traits to biologically credible modelling hypotheses. He knew that the development of coffee beans had three potential outcomes: 1) early failure; 2) late failure; and 3) normal beans. This made it possible to define a probabilistic model that would highlight certain aspects that a coffee grower could easily detect. To do this, de Reffye defined what he called the fertility of a given plant: this was the percentage of ovules that a plant transformed into beans. Using observable genetic characteristics, he hoped to be able to predict, or at the very least evaluate, this fertility as the fructification progressed. The fertility would therefore have to be expressed as a statistical function of these characteristics. Since these characteristics were not continuous, i.e., they were discrete (like those highlighted by Mendel in his time), because they could be reduced to the presence or absence of certain properties (hence the importance of also keeping to a global and practical approach, i.e., a partially *top-down* approach, in order to be able to work on a scale where the biological traits can be harmlessly discretized), it was possible to suggest simple elementary probabilistic laws that, a priori, seemed likely to be responsible for these traits and their multiple combinations.

It was here that de Reffye's approach also became "modellistic", one might say, or synthetic and no longer purely statistical and analytical, since it proposed a scenario governing the creation of observable phenomena in the plant. This scenario was not intended to immediately explain the physiological processes from a causal point of view, but only to allow these observable phenomena to be integrated into an underlying statistical narrative that, unlike the suspected physiological processes, was governed by certain simple probabilistic laws, and especially by combinations of these laws, whose probable behaviour could consequently be predicted by computational means. The statistical scenario therefore did not correspond to a model based on linear hypotheses. In order to be able to mathematically recombine the contributions of each of the probable heterogeneous events affecting the flower, de Reffye chose to assume that the statistical distribution of characteristics was an indication of the emergence of simple laws of probability, such as the binomial law,[16] that were identical no matter what plant was involved. This was the first assumption to be treated as a testable hypothesis. Using characteristics whose statistical rates of presence or absence had been observed and measured on the actual plant made it possible to trace back to their probability of occurring. As a consequence, the "modelling" approach made it necessary to adopt an almost objectivistic interpretation of the probabilities,[17] since it was these probabilities that would, in turn, determine the overall fertility of the plant. In other words, de Reffye did not just stop at highlighting and estimating their values. This was why the probabilities necessitated by the hypothetical scenario had to be dealt with almost as objects. Despite the fact that they did not represent a measurement of a concrete physical being, they were treated as measurements of biological objects because they had been included in an algebraic process that was the only means of providing what he was ultimately attempting to evaluate: the plant's fertility.

Philippe de Reffye thus managed to define four parameters, the first three of which could be estimated in the field using representative samples taken from the coffee plants systematically and on precise dates (which were a function of the degree of maturity of the coffee-cherries): P1 was the probability of an ovule producing an endosperm; P2 represented the probability of an ovule's endosperm ripening (production of beans); r was the probability of ripening of the young fruit; and U was the probability of a flower resulting in a young fruit.[18] De Reffye thus demonstrated that when all the key parameters had been taken into account, the total fertility of a coffee plant could be expressed simply as the product of these four parameters or probabilities: $f = U * P1 * r * P2$. The recombination of these successive probabilistic phases was thus easy in algebraic terms. A large part of the work in his 1975 thesis therefore consisted of separately testing the hypotheses of this combined mathematical model using the distributions given by the samples measured on the coffee plants. To do so, De Reffye used significance tests – in particular the $\chi2$ test (a classic test in inferential statistics since the publication of Karl Pearson's work in 1900[19]). All the adjustments he found seemed satisfactory. For P1 and P2, he updated the characteristic values for each species and for each hybrid form of coffee plant. Thus, for de Reffye, it was this binomial law, together with its parameter or probability, *P*, that became a characteristic of the plant itself as a whole. According to him, this demonstrated that the determination of these probabilities was mainly genetic. The "modellistic" approach therefore did not have just a descriptive or purely fictional and detached impact on agronomists, because it provided a meso-scale frame of reference showing how elementary phenomena could be seen in a different way, without necessarily having to understand them microscopically. To the extent that, within the same plant, different branches can be seen to develop the same probabilities P1 and P2 (a phenomenon that cannot be grasped without this frame of reference, i.e., the model), when observation was guided a priori by the model it became possible to suggest the idea of causal processes, if not on a micro-, then at least on a meso-scale (i.e., on the scale of the whole plant), which could ultimately only be determined by genetics. To de Reffye's mind, even if the model took place on a middle (or meso-) scale, it would still be far from fictitious, even from a primary standpoint.

Thus, when it came to writing up his 1975 thesis, de Reffye wished to emphasize the idea that the "modellistic" approach had indirectly brought to light previously unrecognized biological facts. According to de Reffye, the new ability for understanding offered by the "modellistic" type of mathematical formulation made it legitimate to talk of the "acquisition of experimental facts", and in particular of the experimental fact that "the ripening of beans follows a purely binomial type of process based on the genetic independence of the beans inside the coffee cherry".[20] This meant that the possibility of adjusting a model to data a priori by imposing a hypothesis of independence[21] between beans would be proof of that independence. This new experimental fact was therefore expressed right from the start in an a priori modelling hypothesis that was found a posteriori to be significantly corroborated by the data. Because de Reffye considered the significance test to be an effective validation of the binomial model,[22] since his main epistemological reference remained the theorization of mathematical physics, he considered it allowable to immediately assign a significant biological origin (genetic, in this instance) to this probabilistic fact that he considered to be objective because it was measurable, repeatable for a clone, and could be written as a simple mathematical law.

The next part of de Reffye's thesis work involved the systematic use of this new parameter evaluation tool in order to carry out a comparative study of the fertility of various coffee plants under different conditions. It was here that the most theoretical work could be applied to agronomy. It was confirmed, in particular, that certain parameters that were governed solely by the plant's genetics (P1 and P2) could be clearly dissociated from parameters that were also determined by environment and physiology. Accordingly, it was found that "P1 and P2 are genetic parameters that determine a fixed frequency of late ovule failure, thus affecting the plant's economic value".[23]

As a result of a shift in objectivization that was made possible by objectivistic probabilistic modelling, rather than choosing to conceive the starting characteristics, i.e., the random observable variables (such as the production or non-production of cherries or beans), as genetically programmed traits, de Reffye chose instead to conceive the parameters of the laws of probability of those characteristics in this way. This preliminary objectivization of the laws of probability would later be of great assistance when it came to actually undertaking the integral modelling of the plant's habit.

Nevertheless, when de Reffye defended his first PhD thesis, the examiners were both admiring and at the same time perplexed: was this work based in biology or in mathematics? They recognized that the unclassifiable nature of the work was largely mitigated by its formidable effectiveness in the field. Indeed, it had enabled the best Arabusta to be selected so as to produce a weight equivalent to that produced by the best Robusta (2.6 tonnes/ha), and of superior quality. In reality, the examiners were disturbed by a more epistemological aspect. The work did not consist of settling for a phenomenological approach of an informational type, in the spirit of R.A. Fisher's work, and culminating with variance reductions or principal component analyses between complex phenomena. Instead it claimed a priori to discern a suitable scale – which was admittedly unusual for physiologists, to the contrary of agronomists – on which it could be considered that the phenomena had a sufficiently simple behaviour to allow the usual but combined probabilistic model to be adjusted. Ultimately, this question of scale, or more precisely the originality of the biological level that de Reffye chose to translate mathematically, resulted in indecision for the biologists and statisticians on the examiners' board. For them, the thesis was neither a mathematical theorization of biological phenomena that could be used for purely conceptual purposes, nor a statistical experiment aimed at analysing complex phenomena (experimentation *by means of* a statistical model[24]). They might not have known how to define the work but they knew what it did, and that was the main thing. De Reffye himself publicly acknowledged its pragmatic nature, even though, by his account, the characteristics that he had identified were already what he called "laws of nature", even though they were only valid on a restricted scale.

Nor would this methodological and epistemological decision in favour of a graduated objectivistic modelling, on a suitable scale, be fully understood by his IFCC colleagues. In the early 1970s, in fact, the modelling approach did not appear to be widely accepted among the IFCC agronomists, even though, on the contrary, they quite often used statistical analysis, as well as the British and American biometric methods. Nonetheless, they often recognized that the classic Fisher style of designing experiments was not able to eliminate the extremely variable nature of plantations in tropical environments (such as cocoa farms). But at the time, the IFCC biometricians were trying above all to homogenize their experimental supports *so that the traditional statistical models would apply*, since such homogeneity was required by Fisher-type randomization.[25] The IFCC biometricians in Cameroon, who specialized in cocoa plantations, therefore clearly chose either to find easy-to-manipulate models (construction of abstractions) or to "reduce all variation factors as far as possible, except for those factors whose effects they wished to measure"[26] (data analysis), but they did not choose to *synthesize data*. This was the route that de Reffye ultimately chose, however, as we shall see in more detail.

Fragmented modelling and geometric simulation: de Reffye (1975–1981)

From 1975 (onwards), the search continued at IFCC for models aimed at improving the control and prediction of coffee-plant fructification. The plants' yield could not be entirely explained by the fertility of their flowers. At that point, the plan to observe and model fruit production in detail had just been set in motion with the decision to reify probability, as it were, so as to make it into an observable and quantifiable genetic trait. To de Reffye's mind, it was then necessary to switch from synthesizing the bean to synthesizing the coffee plant as a whole. According to his definition, a plant's "fertility" was simply the percentage of transformation of its ovules into beans. There was a second factor, however, in determining a plant's production of coffee beans: this was the "production capacity" of the plant in producing the flowers themselves. Thus we have: "yield = number of plants/ ha × number of fruits/plant × number of beans/fruit".[27] The number of plants per hectare was known, stable and verified. This variable was contingent on the coffee farmer's informed choices. The number of beans per fruit, in its turn, could be predicted by probabilistic binomial models specific to each clone or each hybrid, thanks to de Reffye's earlier work – and more specifically his first PhD thesis. *But what remained very difficult to evaluate was the number of fruits (or flowers) per tree.* "The numerous variables that affect production capacity make it particularly complex to analyse".[28] It seemed that this factor could not be synthesized directly using elementary modelling scenarios and simple multiplicative recombination, as had been the case for fertility. In order to evaluate the plant's "production capacity", it was necessary to concentrate instead on its morphological aspects: its branching, branch morphology, type of growth, etc. Furthermore, a tool was required in order to consider all these details together, otherwise it would not be possible to calculate the number of fruits per plant. What was required was to study the plant and its morphology as a whole. It was therefore necessary to consider "all the plant's architectural and growth characteristics at the same time".[29] In this specific case, since the number of uncontrolled and crossed factors was very large, the work appeared better adapted to biometric methods. But de Reffye was critical of classic biometry: "such a method requires powerful calculation capacities, but for the most part offers only doubtful effectiveness".[30]

The heart of the problem lay in the fact that, since multivariate analysis made no a priori choices, the questions it put to nature were too open: as a result, it reaped the worst along with the best, but without always being able to organize either into practical and immediately operative knowledge. It was precisely in this sense that the empiricism of biometrics paradoxically verged on speculation, especially in agronomy. On the contrary, although modelling seemed more theoretical in certain respects because of its greater deductive bias, it addressed more closed questions to nature as a result of this a priori basis. In this way, modelling could be more functional. It was the preciseness of its answers and its decisiveness, independently of whether or not it used the laws of probability, that gave it its foothold in the field. De Reffye considered these general arguments to be even more decisive because the purely inductive-type uses of multivariate analysis in fact gave no useable result for the precise problem that interested him. His initial epistemological choice (seeking mathematically expressible "laws of nature") was therefore also strengthened by his research approach. These laws were in fact his models.

Lastly, de Reffye perceived another drawback in using multivariate analysis for questions of yield in agronomy. As we will see, this criticism was essential because it acknowledged a limitation of multivariate approaches that would ultimately justify de Reffye's decision to turn to simulation, i.e., to transition from data analysis to the synthesis of objects, in this specific case synthesizing the growth and architecture of coffee plants. De Reffye's criticism can be condensed into one simple sentence: multivariate analysis was guilty of disregarding a certain number of data. In fact, even if multivariate analysis did not appear a priori to overlook any data, the measurements used corresponded only to a specific instant of measurement, whereas the plant was in fact constantly changing. The plants were therefore directly compared while ignoring their specific histories. The difference between certain *types of growth* therefore was not shown. But de Reffye's text did not dwell on this issue of temporality, which to his mind was obvious, even though at a superficial glance it might appear to be the fundamental reason why dynamic approaches, i.e., simulation, were generally chosen. In effect, de Reffye was also criticizing how the spatiality of arborescent phenomena was dealt with. His complaint that spatiality had been overlooked was, in fact, an intrinsic part of his criticism regarding temporality: the one cannot exist without the other.

> In fact, the measurements are only valid for the instant they are taken, because the plant is constantly evolving. These sets of measurements are usually processed using multivariable statistical methods in order to study the variability of the material. But multivariable analysis does not allow for direct visualization of the architecture because the latter is concentrated in a single point. The differences between two plants can only be expressed in terms of a distance between two points, resulting in considerable loss of information. The results acquired are therefore always rather constrained.[31]

In multivariate analysis, the plant is in fact represented at a given instant by a point in the middle of a cloud of other points that represent the states of the other plants. It is this cloud of points that is studied analytically, in particular by variance analyses that involve minimizing the distances in this multidimensional space. Analysing and minimizing the variance entails, in particular, finding the axes of inertia of this cloud of points and then expressing the data in this new frame of reference. According to de Reffye, it was this homogeneous distance between points, which was both instantaneous and constructed in an abstract space, that resulted in the loss of information. It was being compared against things that were not comparable. On the contrary, it was necessary that dynamized time should take the differentiated space into account. This homogenizing and abstract distance distorted the complexity of the spatial and temporal phenomenon by obscuring it. In certain cases, it was necessary to reject multivariate analysis and its ability to abstract and condense in order to retain an awareness of the variability of living phenomena. The plant's architecture and growth would therefore have to be represented in a more visual and less abstract manner. It was at this point that "visualization" was proposed as an alternative to representation by points and abstract geometrical distances, i.e., as an alternative to the abstractive condensing that results from statistical analysis.

A weakening of the traditional allometric relationships appeared at the same time as the limitations of multivariate analysis. In order to confirm this point, which in fact was already apparent in his first work in 1974, de Reffye dug out an article from 1939 by the horticulturist and botanist J. Herb Beaumont.[32] He hoped the article would further justify the necessity of making the representation of the plant heterogeneous. In effect, Beaumont demonstrated that a simple allometry did not exist for the fruit, but that it was possible, within certain limits, to predict a harvest by using the overall tree growth, taking into account the preceding year's harvest. In his data table, he contrasted the tree's annual growth (calculated as number of branches per tree) with the cross-section of the vertical branches. The latter measurement, however, proved to be irrelevant as it correlated minimally with the annual production of fruit. As in de Reffye's case later, this idea of considering a priori the metric traits (height, size) of the vegetative organs was naturally suggested to him by the research on plant allometry that had been widely reported in horticultural and agronomic work from the mid-1920s onwards, i.e., since the publication of the works of Julian Huxley and Georges Teissier.[33] In the end, it was this negative result that de Reffye found in 1974, when he was still unaware of Beaumont's work. By 1976 it had become clear that it was more important to refer initially to the morphology of the coffee plant than to its physiology, to its increase in length (primary growth) and its structure rather than to its growth in thickness (secondary growth). Thus, in the initial stages, just an estimation of the increase in number of branches would suffice in order to estimate the yield of coffee cherries of the year in progress.

Thus, ultimately, it was for these two reasons (the loss of information in variance analyses and the loss of impetus in allometric models) that de Reffye opted for what might be called an epistemological rift. He decided to devise a dynamic formal representation that would first take the form of a continuous model of coffee-plant growth. Since there was no simple criterion that could predict plant behaviour, the only valid marker was clearly the plant itself, as a whole, seen from the point of view of its morphological history. It was necessary to consider the entire life of the plant, the very history of its morphogenesis. From that point onwards, the model should no longer rely solely on the standardized abstractive tools of statistical analysis that are based on a deliberate condensing of information. The question then was: what type of modelling would his epistemology, such as the technical limitations he had identified, lead him to? It was from this point that almost all his subsequent work became progressively predominated by a modelling technique, which I propose to call "fragmented" modelling, backed up by a simulation and visualization technique that was often of photorealistic quality.

In this case, the model had to be able to predict, for each future instant, the number of branches that would form. Only then would it be possible to extract the quantity of fruit-bearing branches from this number. In the coffee plant, the main stem growing vertically (called the "orthotropic" stem) creates levels of branches that, in their turn, grow in a distinctly horizontal direction (these are known as "plagiotropic" levels). De Reffye, however, based himself above all on the fact – well known in botany – that it is possible to distinguish the branch's creation stage from its growth stage. In accordance with this important consideration, he considered it would be possible to make the model estimate the total number of plagiotropic nodes on a plant at any given instant. By using this dissociation between the process of formation of the plagiotropic levels and their growth process, the mathematical modelling procedure could be simplified since it could be broken down into two steps, each of which could be modelled more simply. It should be noted that it was not the modelling itself that was simplified here, in the sense that the model would become reductive; merely that the modelling was simpler to carry out. The model would be reductive if it disregarded a large number of details. In fact, when we say that such a model is simple, we mean that it is simpler to create in situations when it chooses to conveniently conform to the real dissociation that appears to take place between the various processes affecting the actual phenomena (where these phenomena are observed and identified by means of a common and stabilized empirical procedure, i.e., without recourse to either abstract or fictitious interme-diary entities). Thus what was being simplified in this case, unlike in the classic biometrical approach, was the procedure of creating the model, but not the overall resulting model, nor even its representation of reality.

This was where the computer came into play, because in 1976 it became possible for a complete model to be created, step by step, without having to be drawn up first on paper in its entire final formulation (like the 1975 mathematical model had been), when an automatic programmable digital computer became available to agronomic modellers. This *step-by-step* simplification of the model's preparation procedure resulted in a more complex model, but one that was capable of being supported by the computer infrastructure. The availability of a new, more powerful and program-mable calculation tool thus contributed to the deployment of a type of modelling based mainly on spatial morphology (the structure), unlike the traditional statistical approaches that centred on physiology (the functioning). The digital computer was thus able to oversee the *step-by-step recombination* of sub-models that the scientists conceived separately, in a disconnected or fragmented way. Thus, from having been mathematical, the model was able to become logical and mathematical.

Essentially, de Reffye demonstrated that it was possible first of all to estimate the average time interval, ΔT, required for the formation of a plagiotropic level at any given instant, T. Once this level had been formed, it then became pos-sible in a second phase (for both the calculation and the program) to estimate the parameters of the growth function of a standard branch at that level. Thus, by combining these two modelling functions by means of automatic calcula-tion and using conditional branch logic processed by computer language (which was new with respect to 1975, and which removed the completely mathematical nature of the model) the total number of plagiotropic nodes present on a plant at a given instant T could be determined. De Reffye demonstrated first that an approach using average values of the parameters of the fragmentary events could be retained. For the sub-model of branch formation, de Reffye and his colleague J. Snoeck, the Director of the Genetics Department, were able to demonstrate that it was possible to adjust the curve $F(T) = \Delta N/\Delta T = K \cdot (T)^p \cdot (T_0 - T)^q$, where $\Delta N/\Delta T$ is the monthly increase in number of plagiotropic levels. In this equation, K, p and q are unknown coefficients, but which must be estimated by adjustment, i.e., by using empirical curves. This was possible by simple linear regression once the above equation had been linearized in the following way (assuming here that

$\Delta T = 1$ month): Log $\Delta N =$ Log $K + p$ Log $T + q$ Log $(T_0 - T)$. The total number of levels that the coffee plant might form could then be explicitly expressed as:

$$N_0 = \frac{K(T_0)^{p+q+1} * p! * q!}{(p+q+1)!}$$

This type of analytical formulation was of no immediate interest, however, since it was necessary to be able to precisely determine the age of each of the plagiotropic-level nodes in order to apply to each its own mathematical growth function. Thus, if the aim was to evaluate the number of fruit-bearing nodes on a plant at a given age, it would be necessary to take it to an even more detailed level. The global equation only made it possible to determine the number of levels created each month at a given age. But it was the total number of plagiotropic levels at *an established age* that must be discovered, in order to then be able to make them grow one by one, starting from their date of appearance. It was therefore necessary to proceed *step by step*, first expressing the time lapse necessary for a new level to appear at a given age, and then adding up all the elementary time lapses that were valid at each instant, $T(i)$, until the coffee-plant age, T, could be obtained by progressive summation: $T = \Sigma \, \Delta T(i) = \Sigma \, F^{-1} \, (T(i))$. It was necessary, therefore, to take the inverse $(F^{-1}(T))$ of the preceding function $F(T)$, and to add all the numerical values of this function at each of the different instants when a new level appeared. It was precisely this type of iterative calculation that began to make the use of a digital computer so necessary: "The combining of all the ΔT values can only be effected with a digital computer".[34] The calculation of the orthotropic growth was therefore broken down and carried out step by step, since there was no longer a simple and explicit mathematical equation. The computer was essential here, because it made it possible to calculate, for each instant, the lapse of time that would be necessary for another event (i.e., the formation of a further plagiotropic level) to occur, right up to the age T under study. In so doing, the computer program stacked in one variable, N ($N = N + 1$ with each iteration), the number of times that it had carried out this calculation. The number N obtained as a result, i.e., when the program had finished, corresponded perfectly to the total number of plagiotropic levels of the coffee plant at age T.

Next, for the branch-growth sub-model, by drawing the curve of number of nodes that had grown as a function of time, de Reffye and Snoeck found that it could be modelled realistically by the simple mathematical model $g(t) = n_0 \, (1 - e^{-rt})$, where n_0 was the maximum number of nodes that a plagiotrope could reach, and r was a parameter measuring speed of growth.[35] The only justification they offered for this was expressed in their assertion that it was – in their own words – the shape of the observed curve that had "suggested"[36] this type of model to them. Likewise, by linearizing in the same way as earlier, i.e., by a transition to the use of logarithms, they were able by means of regression to find an adjustment that turned out to be "excellent", in their view. In this way, they found, for "clone 182" of *Coffea robusta*, a function $g(t)$ that expressed the number of nodes produced per unit of time on a given plagiotropic level. But not all the nodes that formed were fruit-bearing. For each instant, it was therefore necessary to be able to determine how many of the nodes that formed were actually fruit-bearing.

During the next step, the authors applied a certain level of botanical knowledge and observation, as they had done for the hypothesis of dissociation between the formation and growth processes of the plagiotropic levels. In effect, it can be considered that, after their formation and during their growth, the plagiotropic levels go through three distinct successive phases: the nodes that have formed produce leaves; they then enter a phase during which they bear fruit; and finally they shed their leaves. On a single branch there are therefore three zones: a zone with leafy nodes at the tip of the branch; a zone with leafless nodes at the base of the branch; and a zone with fruit-bearing nodes located between the two. According to the records that were available, however, it was possible to estimate the lifespan of a leaf, as well as that of a fruit-bearing node. Any node that appeared at the tip of the branch would at first display the presence of leaves. It would remain this way until a given time, t_0, which is the average lifespan of a leaf. Thus, on a given plagiotropic branch, the total number of nodes with leaves at instant t is:

$$\text{if } t \leq t_0: \quad n_{leaves} = g(t)$$

$$\text{if } t > t_0: \quad n_{leaves} = g(t) - g(t - t_0)$$

Then, since the fruit-bearing nodes are "equally restricted in time between a minimum time of appearance, t_1, and a maximum lifespan, t_2," "there are three possibilities regarding a branch at level k, with an age of $T - \sum_1^k \Delta T_i = t$":[37]

$$\text{if } t < t_1: \quad n_{fruit\text{-}bearing} = 0 \text{ (the branch bears only leaves)}$$

$$\text{if } t_2 > t > t_1: \quad n_{fruit\text{-}bearing} = g(t - t_1)$$

$$\text{if } t > t_2 > t_1: \quad n_{fruit\text{-}bearing} = g(t - t_1) - g(t - t_2)$$

According to the observations on actual plants, the authors estimated that on average: $t_0 = 10$ months, $t_1 = 6$ months and $t_2 = 19$ months.[38]

They were thus able to give these equations in numbers. In the end, all the partial mathematical models were interpreted quantitatively, and in 1976 the authors were able to propose what they called a "Global synthesis of coffee-plant growth", in the form of a program setting out this continuous model in five steps.

1 At the start of the program, the final age, T, of the plant is entered.
2 Branch formation: using the function, $F(t)$, the program calculates the time necessary for the first branch or the next branch to appear. The current age, t, of the plant is then incremented by this time lapse. The age of the current branch will therefore be $T - t$.
3 State of growth of the branch: two solutions are then possible: if $T - t < t_1$, the branch is too young and bears only leaves, and the program halts because any other branch that grew would be even younger, and therefore would also not

bear fruit; otherwise, there are also two other sub-possibilities: if $T - t < t_2$, then $\Delta n_{fruit\text{-}bearing} = g(T - t - t_1)$ (the branch is too young to bear leafless nodes), otherwise $\Delta n_{fruit\text{-}bearing} = g(T - t - t_1) - g(T - t - t_2)$[39] (these are branches that are neither young nor old, and that therefore have both fruit-bearing nodes and nodes that have shed their leaves).

4 Next, the program increments the current number of nodes of $\Delta n_{fruit\text{-}bearing}$ calculated earlier, and increments the current number of branches by 1 (end of the test of the branch's state of growth).

5 Finally, if $t < T$, the program loops back to step 2, which will deal with the formation of the next branch (end of the test of current branch's formation), and otherwise it stops.

At the end, the current number of branches is the total number of branches on the plant at age T and the current number of fruit-bearing nodes is the total number of its fruit-bearing nodes.

Using this flowchart, it can be clearly seen that the computer processing of the first mathematical model, $F(t)$, is sampled, or fragmented, so as to allow each of the steps determined by the function $F(t)$ to carry out the processing of the second mathematical model, $g(t)$. Without computer programming,[40] it would not have been possible to carry out this fragmentation and so to keep the complexity of each of these mathematical models unchanged. It would have been necessary to find simpler models whose composition could be calculated by hand, or else a global mathematical model that could also be calculated by hand, like those normally proposed by the IFCC biometricians.

De Reffye later added a further modular element (or sub-model) that specifically took into account the mechanical properties of the branches, and that would address, step by step, the issue of lodging or shoot bending under self-weight loading[41] and breakage. He therefore further increased the realism of the morphogenetic coffee-plant model, and this realism would be confirmed by means of the plotter drawings of simulated coffee plants. Thus, in the end, there were four principal modular elements, each corresponding to a sub-model that was specific to a particular biological phenomenon: stem growth, branch growth, stem lodging and branch bending. These four mathematical sub-models could be used in conjunction with each other: they exploited each other successively and mutually according to logical conditions determined by the quantitative results of the sub-models themselves (logical branch conditions centred on critical time exceedances or on critical branch weight, etc.). The structure of the automaton, or the sequence of logical states, of such a program (which was nonetheless very simple) therefore depended on the functioning of the automaton itself from the point of view of its mathematical and logical aspects. The step-by-step functioning was therefore indispensable in order to know its structure and dynamics. This was precisely one of the major breaks with mathematical modelling, whether theoretical or practical. From that point onwards, de Reffye would talk of simulation rather than of modelling. Yet simulation and visualization were not initially presented as ends in themselves. It was fragmented and calibrated modelling that was primarily sought in the beginning. Simulation and visualization therefore had instead to play the argumentative role of a second empirical confirmation, since they made it possible to directly compare the "theoretical" results with the phenomena observable in the field: by a sort of transitivity of confirmation, the experiment confirmed the simulation, which in turn confirmed the sub-model, which in this instance was theoretical and explanatory (as far as the phenomenon of lodging or shoot bending under self-weight loading was concerned), since it was based on the theory of resistance of materials. This comparison between the field measurements and the simulation could be made by eye, but could also be carried out more rigorously using suitable statistical tools. Indeed, thanks to the technique of fragmented and recombined mathematical modelling, the visualization of the lodging or shoot bending was based on a theoretical formalization that appeared clearly to be added on to and intertwined with the earlier formalization of architectural and geometric growth; this formalization, in turn, was primarily descriptive and phenomenological, but had already been validated (in the sense of calibrated, in this case) in its own right.

In the end, it turned out that reconstitution by means of sub-models based on the theory of resistance of materials (in particular with "critical load" formulas to determine the force at which the coffee-plant stem "buckles"), together with a scenario combining the genetic traits implicated in breakage, could be used to designate the genetic markers that must ultimately be selected. The logical and mathematical modelling that de Reffye recommended gave the experimenter the means to measure precisely and without ambiguity the main characteristics involved in the equations of the mechanics-based sub-model (Young's modulus, internode length, stem diameter). Lastly, the model gave the breed-selector the possibility of having access to precise criteria to make their decisions. The selector could thus measure criteria and at the same time predict the stability behaviour of the selected clones.

This type of logical and mathematical modelling may well be a combined formalization, or, we might say, impure, but nonetheless its decisive advantage was that it could designate and integrate the "experimentable", i.e., in this case, that which can be measured, in predictive calculations. The modelling therefore did not remain at the epistemic level of a preliminary model calculation that cannot be calculated analytically, nor at the level of a mere test for a theoretical view, since it referred precisely to a view that, as we have seen, was by then fragmented and far from united (when field biology was at issue). Given this situation, de Reffye considered that it was a computer simulation and no longer even a model with two conditions: on the one hand, it can be said that we *simulate* on a digital computer from the moment a *mixed model*, that is at least logical and mathematical and *conceived in a fragmented manner* (even if these were not the terms he used), becomes the basis of *step-by-step calculations* that successively bring into play *different models* (sub-models) of different parts and different properties of the organism. On the other hand, according to de Reffye, there is simulation if there is an *ability to visualize or to measure the results of these calculations in one way or another*.

There are thus clearly at least two phases in a simulation: *replicatory interactions and measurement (or observation)*. In this case, the visual result presented by the computer and plotter closely resembles the phenomenon perceived by our own eyes and on our own level: it is therefore justifiable to speak of feigning, of simulating. The emphasis is thus on simulation and not on the global model that was intended to make it possible, but which no longer exists as a unified model outside of its computer implementation.[42]

This important work was published in the IFCC journal (*Café, Cacao, Thé*), but evoked very little reaction at first, other than for the immediate uses of his plant-variety selector colleagues at Bingerville. Revealingly, in the view of the editorial committee that classified the articles at the end of each year, it was not the method, the technique or the modelling itself that should be stressed but rather the phenomenon that had been modelled (i.e., the lodging or shoot bending) and the agronomic problem that was addressed. The proposed method's potential was therefore far from being perceived at that time. There is one more point that should be noted in order to conclude this introduction to the fragmented modelling and simulation technique. De Reffye produced this work in considerable isolation. He therefore did not cite the works of Cohen or Honda, even though they had already proposed their geometric and graphic approaches to the simulation of branching shapes. He was therefore in no way influenced by their work. In fact, his computer equipment was much less powerful, and he would not have been able to build on their work in exactly the same direction. Unlike Cohen and Honda, de Reffye placed greater stress from the outset on those morphological characteristics – the nodes – that were so important in agronomy: he was not interested in their aesthetic visual appearance for the sake of theoretical argument: his aim was to arrive at a precise quantification of the number of fruit-bearing nodes. He was therefore obliged to draw on much more precise botanical knowledge than that used by Cohen, and later by Honda and Fisher.

Simulation, imitation and the sub-symbolic use of formalisms

In his subsequent work, in addition to the first two conditions determining the use of the term "simulation" that he had already chosen (i.e., fragmentation for replication and graphic visualization), de Reffye would add other conditions that had already been used for some time to regulate the use of the same term in other fields, such as nuclear physics (these conditions appear in Eden and Cohen, for example). In the context of a new set of agronomic problems, de Reffye once again found inspiration in operational research through one of its other methods of descriptive mathematics: probabilistic simulation. In the years that followed, thanks to a significant work on cocoa-tree pollination by different insects (which there is no point in describing in detail here, since it exceeds the scope of plant morphogenesis), de Reffye made further considerable developments to his fragmented modelling and simulation approach. Basing his work directly on the seminal work by Thomas Naylor and Joseph Balintfy, he decisively incorporated the Monte-Carlo method[43] in a simulation of the individual behaviour of pollinating insects. It was for this reason that he was also able to use probabilistic simulation as an empirical test for certain analytical sub-models. For de Reffye, such probabilistic simulation unquestionably formed a part of experimentation, but from a different aspect than those already mentioned in the case of the first visualization simulations. In a 1977 article, he dwelt in particular on the advantage of having the ability to integrally reconstruct the distributions of random events and not just their averages or their variances. According to de Reffye, simulation allowed "verification of the accuracy of the mathematical analysis".[44] It was in this respect that simulation could be considered an experiment: it was able to corroborate a theoretical view, but – to the extent that it was a *step-by-step* reconstruction of the real phenomenon – it was a reconstruction that was itself nonetheless underpinned by the theoretical hypothesis. This was not a circular argument because it meant, more precisely, that the simulation was able to corroborate a novel two-level theoretical framework, namely a *theoretical break-down* (fragmentation) of *partial theoretical views* (each one implemented via modular elements that were specialized in branching processes, fruit- and leaf-formation, leaf-shedding or wood mechanics). Indeed, starting from a logical and mathematical representation, it was possible to compare the simulated results with the theoretical ones or, in other words, to compare two results that were products of the same initial logical and mathematical representation, but that did not originate from the same interpretation of that representation (i.e., condensed and modelled probabilistic laws vs event-generating probabilistic laws that are simulated step by step). Insofar as the single-level "theoretical results" were concerned, it was an interpretation that was primarily abstractive and condensing because it was centred on the averages of the phenomena and on the parameters of the mathematical models (where neither time nor the effective diversity of types of events were considered). The "simulated results", on the contrary, were a constructive interpretation of those same models, coupled with a regenerative use regarding their temporal and individual events. In this case, simulated restitution was less abstract than the abstractive mathematical model since it retained and used at least one of the so-called concrete dimensions of the initial phenomenon as support for its manifestation – time – resulting in what may be called a weakening of the formalism's symbolic condensing property. The representation encounters the phenomenon and "touches it", one might say, in this dimension at any rate. In this way, the simulation derived from capturing rather than from condensing the empirical. The similarity in aspect (if the temporality of a phenomenon is considered to be one of its aspects) was therefore greater in the stochastic simulation than in the abstractive interpretation of a mathematical modelling, notably when time was excluded: the abstractive transfiguration of the empirical was less in this case.

Nevertheless, this specific similarity of aspect (i.e., the dynamic resemblance) was not necessary in order to have a simulation. While many authors,[45] following similar but more generalized arguments, have held that dynamic resemblance (which can also be called trajectory imitation) is key, and effectively characterizes all simulation, I have in fact already demonstrated that this point of view – which we can see applies above all to simulation by stochastic processes (or more broadly by path-dependent calculations) – can be put into perspective, and that de Reffye's work clearly introduced, along with dynamic resemblance, many other types of existing simulations. On the one hand, many mathematized laws possess this dynamic property of temporal resemblance or "over time". On the other hand, we have seen that there are reasons to already give the name "simulation" to the construction of models that produce resemblances that are not essentially dynamic, but rather are visual. Resemblance *by* dynamic should be distinguished from resemblance *of* dynamics. The latter is more constraining and is not necessary for every simulation. But for that matter, can resemblance by dynamics itself be defined or constructed without relying on another type of resemblance, or at least a correspondence of terms (even if only for the purposes of defining an order of calculation), and which in that regard precedes it? We will see later that de Reffye's team was able to simulate tree architectures on the computer by means of temporally dissimilar step-by-step simulation stages (AMAPsim software), the results of which were nonetheless visually similar. One of the issues at stake was precisely to then try to capture the missing aspect of the resemblance: a tree that would be botanically, i.e., spatially, realistic at each moment of calculation time, when the calculation time itself was realistically ordered. In this sense, there may therefore be spatial resemblance broadly speaking (i.e., partial replication), without temporal resemblance, i.e., similarity of sequencing. The contrary appears more open to doubt. This suggests in all cases that, with the frequent characterization of any simulation as a "model in time", we have not yet found the real characteristics that might give unity to the concept of simulation, in the way it is implemented, deployed or extended by modern science, and in particular by computational science. Although it still remains to be proven, it is likely that the notions of dilation or of dilated representation (taking the more explicit opposing view of classic symbolic, mathematical and logical condensation to a more transverse level) as well as the notion of selective desymbolization, could become more generally applicable in their current state.[46]

The fact remains that, if we retrospectively analyse de Reffye's way of working, we can see that it was the precision and urgency of his pragmatically based questions that led to his engineering-type approach, contrary to that of Hisao Honda, for example. For that matter, his view of experimental science and of the role that formalisms should play in it spurred him, in his work on plant-growth improvement, to focus his attention and the attention of his models on the processes themselves of growth and fruit-bearing, rather than focusing – as scientists working solely in biometrics or in data analysis had done until then – on strict *ex post* plant-variety selection (i.e., selection that was effected after the event, based solely on the overall plant performance). In light of the very specific problems he faced at the start, he was therefore driven to produce operational modelling solutions that in fact outperformed purely data-driven and case-by-case selection. As we will see, it was primarily in his second doctoral thesis, and later in interacting with more experienced botanists, that de Reffye would become truly aware of the potential that his approach could offer in achieving a more fundamental biological science through increased universality, particularly insofar as the representation of plants in general and of their growth dynamics were concerned.

The new question that would justify this thesis work came to him as a result of his dissatisfaction with his fragmented but kinetic and deterministic growth model of 1976, even though this model already met many of the requirements of his colleagues, including his own immediate supervisor, J. Snoeck. In actual coffee plants, in fact, a fairly high degree of irregularity can be observed, or more precisely a variability in node life spans, irrespective of node type. Indeed, the coffee plant is one of the plants that best displays this type of irregularity. Hence the suggestion that he should take this variability into consideration in order to produce a computer representation that would be more architecturally faithful to the actual individual cases seen in the field. De Reffye did not forget that, even though he had simulated the plants' growth in a chronological manner (and therefore realistically in the sense that it was "ordered" in this temporal dimension), he had in fact only considered the average life spans of the events that successively affected the plant's nodes. The trajectory imitation of this first simulation was therefore only valid as an average, for each time step. By chance, it so happened that this first approach by averages was sufficient for, and was even very well suited to, the case under study since, given the stability of Côte d'Ivoire's subtropical climate, their coffee plants generally displayed significant continuity in average growth, despite a strong tendency towards dispersion around the mean values. But since de Reffye had, for that matter, become more familiar in the meantime with stochastic processes, especially through his work on cocoa-tree pollination, it seemed natural and appealing to him to try to increase the complexity of his architectural representation of coffee-plant growth by using these same processes to make it even more realistic from a botanical point of view. In 1978, this new work on a stochastic modelling of coffee-plant architecture was already well under way when de Reffye finally spoke about it to Yves Demarly (born 1927), a Professor in Applied Genetics and Plant Improvement at the University of Orsay, who had been supervisor for de Reffye's first PhD thesis and had agreed to also supervise his second PhD thesis. In Demarly, de Reffye found a researcher with an exceptionally open mind, who was particularly unlikely to curb an adventurous spirit such as his own.

Nonetheless, in 1978 his draft thesis would be given an even more generalized foundation following a very specific event. This was a meeting with the botanist Francis Hallé (born 1938), who would ultimately steer him decisively from his initial simulation of just the architectural growth of the coffee plant towards a very opportune extension that would include simulation of the growth of almost all known plants, in view of a new concept that had just been proposed by Francis Hallé himself and by his Dutch colleague Roelof Arent Albert Oldeman (born 1937): this was the concept of "architectural model". It should be understood that, by promising to work towards greater botanical accuracy, de Reffye's research would normally be faced with completely different scientific traditions than those of quantitative agronomy. As we will see later, by following this decisive turning-point towards realistic but not immediately pragmatic modelling, de Reffye would have to situate his research in relation to other, older, more established works, whether these dealt with botany, mathematical biology or of course quantitative morphology, but also in relation to works on computer-generated plant images using computer graphics. Even though de Reffye did not, in fact, have the benefit of an in-depth knowledge of these other, rather theoretical traditions (indeed far from it), his work – perhaps by chance, but also and especially because his initial set of problems (which were agronomic and operational in nature) had not been the same as those that characterized the other traditions – would not be completely assimilated into the others. Since I have already highlighted the status that each of these traditions accorded to the computer (apart from the computer graphics approaches, which we will return to in due time, and which developed in parallel and in direct competition with de Reffye's approach, before ultimately converging with it), we will be more able to understand what de Reffye's approach had in common with these other approaches, as well as what made his approach unique. It will therefore be necessary to grasp what gave this simulation its unprecedented ability to bridge the gap between empirical and practical disciplines and theoretical and descriptive ones. The solely calculative and theoretical use that simulation had primarily had until then in the sphere of plants would therefore be abandoned. Instead, simulation would take a definitive place in fieldwork. By measuring simulation against fieldwork in all its complexity, by grasping the singularities of the field, simulation might even be able to pass for what might be called a "second-type" experiment, where the first type refers to classic experimentation.[47]

But first let us return to the meaning of this botanical revolution involving the introduction of the concept of "architectural model". For it was this revolution that would encourage de Reffye to persevere with the simulation of growth, bud by bud, and that would continue to put him in an awkward position not just with respect to the official epistemological line upheld in the schools of theoretical biology (where modelling = excluding, in order to universalize and explain), but also with respect to the stance of practitioners of field models (for whom modelling = selecting a mono-formalized perspective representation for a single use).

Notes

1 The Higher National Engineering School of Agronomy, located in Toulouse; this is one of France's competitive-entry engineering institutions in agronomy.
2 *Diplôme d'études approfondies*, diploma of advanced studies, comparable with a British Master's degree.
3 Bingerville is a suburb of Abidjan, in Côte d'Ivoire.
4 The French Institute of Coffee, Cocoa and other stimulant crops.
5 This service was established in 1955. The ORSTOM itself (*Office de la recherche scientifique et technique outre-mer* – Office of Overseas Scientific and Technical Research) was set up in 1947 in order to carry on from the *Office de la Recherche Scientifique Coloniale* (ORSC, the Office of Colonial Scientific Research), which had been created by the Vichy Government in 1942.
6 Reffye (de) (P.) *et al.*, *Document préparatoire à la revue externe de l'Unité de modélisation des plantes (Preparatory document prior to external review of the Plant Modelling Unit)*, CIRAD, Montpellier, 1996, p. 3. Most of de Reffye's early work was published in the original French and has not been translated. These, and subsequent, quotations have been translated for the purposes of this book. A list of translations of the various French acronyms can also be found at the beginning of the book.
7 Reffye (de) (P.), "La recherche de l'optimum en amélioration des plantes et son application à une descendance F1 de *Coffea arabusta*" [Search for the optimum in plant

improvement and its application to an F1 offspring of *Coffea arabusta*], *Café, Cacao, Thé*, 1974, Vol. 18, No. 3, pp. 167–178.

8 This is a data analysis on numerical variables, or at the very least on scaled variables (i.e., variables that are ordered in relation to each other with constant relative distances). In the same way as correspondence analysis, which focuses on nominal or qualitative variables, principal component analysis consists of seeking the axes of inertia of the point cloud formed by the data, and of expressing them as a function of these new axes or "principal components".

9 In data analysis works these are also called "explanatory models", in contrast to principal component data analysis models that are often called "descriptive", insofar as they are a specific case of principal component analysis. In this case, there is only one single variable to be explained according to all the others. It is interesting to note that the model is considered to be explanatory, or at any rate predictive, in cases when, as in an explicit mathematical equation ($y = f(x)$), one variable is clearly emphasized by being expressed as a linear combination of all the others.

10 Reffye, (de) (P.), art. cit., p. 169.

11 The "naturalness" of plants can in fact be disputed when new plants are "manufactured" in the context of genetic improvement. Since, by means of this process, the plant becomes manifestly and more than ever a human artefact, and is perceived as such, it may be easier to transfer the engineering methods that were previously applied to the management of human businesses to plants.

12 See Glossary.

13 Allometry: see Glossary.

14 Ibid., p. 176. This suggestion was one of the reasons behind the modelling approach of his 1979 thesis.

15 In the 1970s French university students were required to submit two theses: a postgraduate thesis (known in French as the "*troisième cycle*" or "*bac[calaureat] plus 8 [years]*"), which is equivalent to the present-day PhD, and a State thesis (called a "*thèse d'État*" or "*thèse d'habilitation*"), which was often written over a period of many years.

16 A binomial law takes place during repeated tests on complementary events. Thus, if the universe is reduced to two events, when random selection of either of these two events is repeated *n* times, the respective probabilities of these two events, p and $1 - p$, combine in order to form the probabilities of their different combinations. The values of these combined probabilities are thus equal to the subsequent terms of Newton's binomial ($((1 - p) + p)^n$ (hence the name "binomial law"): the overall probability that the event whose elementary probability is p will be repeated k times over n draws is therefore $P_k = C_n^k p^k (1 - p)^{n - k}$.

17 An objectivistic interpretation views statistical measurements as the manifestation of an objective chance occurring in the phenomena themselves. This type of interpretation often goes hand in hand with a "modellistic" perspective according to the historian and epistemologist Ian Hacking. See *The Taming Of Chance*, Cambridge, Cambridge University Press, 1990, p. 98: "the fundamental distinction between 'objective' and 'subjective' in probability – so often put in terms of frequency *vs.* belief – is between modelling and inference". Subjectivistic interpretation attributes these random fluctuations to the impossibility of being able to completely control or completely know the conditions under which these phenomena are measured. A modellistic approach to fructification, however, appears to be able to explain both: it depends on the ontological level emphasized. If the level of the physical and chemical processes is deemed to be more real, i.e., more ontologically based, then the most aggregate level of emergence of fructification characteristics will – under the laws of probability – be affected by any ignorance about these elementary phenomena. In that case, a subjectivistic interpretation of the fructification model would be chosen and there would be no major break with the main trend of French biometry. But if it is primarily the level of the bean and fruit that is ontologized, which may also be justified,

for example, in an agronomic context (this is what is known as practicing "pragmatic" science, although, as can be seen, this qualification is ultimately entirely relative), then it is this bean- and fruit-level that will be deemed to be elementary. At this level, it could equally be considered that we are dealing with an "objective" chance, giving rise to an interpretation that can be qualified in this case as quasi-objective. In this way, the formalism becomes almost re-rooted.

18 In this explanation, I have simplified de Reffye's model insofar as it ought normally to take into account, as he did, the fact that there are two loculi in a coffee cherry and not just one. This consideration would simply further complicate the calculation, however, without actually modifying the general principles used.

19 Gigerenzer (G.), Swijtnik (Z.), Porter (T.), Daston (L.), Beatty (J.), Krüger (L.), *The Empire of Chance: How Probability Changed Science and Everyday Life*, Cambridge, Cambridge University Press, Ideas in context series, 1989; reprint: 1997, p. 79.

20 Reffye (de) (P.), "Formulation mathématique des facteurs de la fertilité dans le genre *Coffea* [Mathematical formulation of fertility factors in the *Coffea* genus]", post-graduate thesis, Université Paris-Sud Orsay, September 1975, p. 31.

21 This is the independence between two neighbouring beans in reacting to the same type of events during ripening.

22 He could have been challenged on this point. It was therefore an entirely conscious decision on his part, because, as many mathematicians have pointed out since the first appearance of significance tests (even if this is still, justifiably, under debate), there is no purely logical reason allowing us to pass from an analysed frequency to a synthesized probability. The objectivity of the measured frequency cannot be compared with the objectivity of the assumed probability on the basis of the measured frequency. This is one of the most general characteristics of mathematical models: if one model is possible, then another model is always equally possible. De Reffye clearly based himself here on a practical certainty, not a logical certainty – i.e., a certainty that came from axiomatized rules that were internal to the statistical tool.

23 Reffye (de) (P.), 1975, op. cit., p. 71.

24 For this conception of the model, see Legay (J.M.), "La méthode des modèles, état actuel de la méthode expérimentale" [The model method, current state of the experimental method], *Informatique et Biosphère* [*Computer Science and Biosphere*], Paris, 1973, pp. 5–73.

25 See Glossary: "experimental design".

26 Lotodé (R.), "Possibilités d'amélioration de l'expérimentation sur cacaoyers" [Possibilities for improvement in experimentation on cocoa trees], *Café, Cacao, Thé*, 1971, Vol. 15, No. 2, April–June, p. 91.

27 Reffye (de) (P.), Snoeck (J.), "Modèle mathématique de base pour l'étude et la simulation de la croissance et de l'architecture du *Coffea robusta* [Basic mathematical model for study and simulation of the growth and architecture of *Coffea robusta*]", *Café, Cacao, Thé*, 1976, Vol. 20, No. 1, p. 11.

28 Ibid.

29 Ibid.

30 Ibid.

31 Ibid.

32 J.H. Beaumont was "Principal Horticulturist" at the Hawaii Agricultural Experiment Station. He was named Director of the Experimental Station in Hawaii in 1936. He would later be recognized throughout Hawaii and the rest of the United States for his 1953 studies on a type of Macadamia nut tree with considerable ornamental qualities, which he subsequently imported from Australia to Hawaii and later to the Californian department of horticultural science. In 1965, following a decision by the California Macadamia Society, this Macadamia variety was named after him.

33 See Gayon (J.), "History of the concept of allometry", *American Zoologist*, 2000, Vol. 40, No. 5, pp. 748–758.

34   Reffye (de) (P.), Snoeck (J.), 1976, art. cit., p. 15.
35   Ibid., p. 18.
36   Ibid., p. 18.
37   Ibid., p. 17. The reminder here about ΔT is intended to prepare us to include the number of branches in the overall flowchart of growth processing, i.e., in the combined processing of the second partial model by the first partial model.
38   Ibid., pp. 18–19.
39   Ibid., p. 23.
40   Before IFCC acquired a similar model, it was the physicists of the Abidjan Faculty who initially loaned de Reffye their brand new Hewlett-Packard HP-9820 A digital computer that had been available on the market since 1972.
41   See Glossary.
42   A particular computer program set-up, linked especially to the specific constraints of the operating system and of the programming language selected.
43   See Glossary.
44   Parvais (J.P.), Reffye (de) (P.), Lucas (P.), "Observations sur la pollinisation libre chez *Theobroma Cacao*: analyse mathématique des données et modélisation" [Observations on open pollination in *Theobroma cacao*: mathematical analysis of data and modelling], *Café, Cacao, Thé*, 1977, Vol. 21, No. 4, October–December, p. 261.
45   This was the opinion of Stephan Hartmann and Paul Humphreys, in particular. Cf. Humphreys (P.), *Extending Ourselves: Computational Science, Empiricism and Scientific Method*, Oxford, Oxford University Press, 2004, p. 108.
46   See Glossary for definitions of "computer simulation" and "sub-symbolic". See the conclusion for a systematic recap of these points.
47   Cf. Varenne (F.), "La simulation conçue comme expérience concrète" [Simulation conceived as a concrete experience], in J.P. Müller, Ed., *Le statut épistémologique de la simulation, Proceedings of the 10th journées de Rochebrune* [The epistemological status of simulation], Paris, Éditions de l'École Nationale Supérieure des Télécommunications de Paris (ENST), 2003, pp. 299–313. For a review of the main arguments of that article, see conclusion.

4 A random and universal architectural simulation

In this chapter we will see how the process of fragmented modelling and integrative computer simulation, which had originally been developed in order to simulate coffee-plant growth, would lead to an architectural simulation of plants that could be described as universal *thanks to its ability to simulate, by simple extension, the entirety of plant architectures observed in nature. In contrast, I will reveal the technical limitations of the more classic theoretical or biometrical formal models, in particular when it comes to grasping extremely composite objects such as plants. Theoretical models in fact pay little heed either to the complexity of the living essence itself (reductionism) or to the evolving and intertwined nature of the optimization function that is meant to follow plant genesis (a reduction to optimization principles that homogenize and de-historicize the scenario, despite its complex interweaving of cellular differentiation and of growth). For its part, statistical biometry requires simple models for a precise usage, without recognizing that it over-constrains its language, whereas it could be more generous with regard to the data without always reducing them to averages, variances and so on. Computer simulation, on the contrary, enables such* generosity. *Although computer simulation, like biometry, has the advantage of not viewing the plant as a theoretical object, it also allows a sort of underlying* theoria *by constructing a sort of multi-dimensional scale drawing as opposed to the perspectives represented by the models. We will see nonetheless that this search for realism was not always understood or well received by modellers of living beings – to the extent that it risked disappearing and falling into oblivion in the early 1980s.*

In 1978, when de Reffye met Francis Hallé, it had already been at least eight years since Hallé and his colleague Oldeman had proposed a significant conceptual advance in botany. At the end of the 1940s the synthetic theory had effectively imposed a return to understanding the plant on a global scale. Since selection operated essentially on the individual's overall architecture and on populations, it was therefore necessary for botany to find an intermediate biological level that would allow all the issues, whether physiological and morphological or ecological and evolutionary, to be integrated and linked together. It was partly in this spirit that, at the end of the 1960s, the work of ORSTOM botanist Francis Hallé and his Dutch colleague Roelof A.A. Oldeman led them to the concept of "architectural model". First, though, let us return briefly to this concept where the term "model" emerged, albeit without referring to a formal construct.

DOI: 10.4324/9781315159904-5

Making headway in botany: the notion of "architectural model" (1966–1978)

Francis Hallé began his career as a student of botanist Georges Mangenot, who was a professor at the Sorbonne and later at the Orsay Faculty. During the 1930s Mangenot had been a colleague and associate of Lucien Plantefol.[1] After the war, Mangenot was appointed Director of the ORSTOM centre at Adiopodoumé in Côte d'Ivoire. He specialized in tropical botany and focused on a new approach that had recently been advanced, by American ecologists in particular, that was known as dynamic ecology. In 1963, through the intercession of Mangenot, Hallé was sent to Adiopodoumé, where he was initially responsible for research at ORSTOM and subsequently became Director of the Abidjan Botanical Institute. At that time, Hallé's employers did not consider that the vegetative plant parts played an important role in their working methods, and plant branching was seen as inessential. A sort of anarchy appeared to reign. Once he was actually on site in the equatorial country, however, Hallé noted that they rarely studied the flowers (which had been the traditional method of plant recognition since the days of Linnaeus): indeed, the flowers were for the most part inaccessible to the observer. In contrast, he observed that the vegetative shape of the plants was extremely clear and very simple: it sufficed to simply sketch them, since in fact they already looked like drawings from the outset.[2] The Ivorians themselves identified plants solely on the basis of their vegetative shape. Hallé therefore devoted himself almost exclusively to observation of plant architectures and rumination on plant shapes. It was Hallé who proposed the term "vegetative architecture" in 1964, and the suggestion was endorsed during discussions with the Orsay botanist René Nozeran. The same year, Oldeman arrived at ORSTOM and began to collaborate with Hallé; their joint work proved to be seminal. It was published in 1970 in a monograph entitled *Essai sur l'architecture et la dynamique de croissance des arbres tropicaux*.[3] This work presented observation and systematic testing of the different types of architecture on the one hand, and on the other hand provided a summary of various works that had already been published. At the same time, they also proposed a clarified and stabilized terminology. Instead of the Goethean term "morphology", the authors opted definitively for the term "architecture", since this term could designate the purely structural morphological characteristics of the plants, rather than other morphological traits such as presence of latex, pilosity or limb thickness, for example. For Hallé and Oldeman, investigations should focus primarily on the plants' external configuration, shape and growth dynamics, which appeared to be determined strictly by genetics. Their method consisted of direct observations in the field, or else growing certain trees so as to observe their morphogenesis as young plants of less than 15 metres in height, since at this stage the "young tree, when protected from ecological trauma, freely expresses its *ideal shape* as dictated by its genetic makeup".[4] After this stage, in effect, an "alteration of the specific organism under the influence of the macroclimate"[5] could be observed: by that point, the tree had been subjected to numerous injuries, with the result that its "statistical habit" no longer resembled the phenotypical habit of its early days. Oldeman had demonstrated in practice that injury or even simple ageing gave rise to what he called "reiterations" of the "architectural model" within the injured or aged plant. These "reiterations" could appear as suckers or secondary trunks. Hallé explained this phenomenon by the fact that, during ageing or in the case of injury, there was a weakening of the "network of morphogenetic stresses".[6]

It is this term, "ideal shape", that best characterizes the spirit in which, immediately afterwards, Hallé and Oldeman then proposed the term "model". This term was used to designate a paradigm, rather like a Platonic idea that they had inherited explicitly from the earlier Goethean speculations on the primitive plant (*Urpflanze*). It therefore had nothing to do with a mathematical model in the positivistic sense of a veneer of formalism over a natural reality. It was primarily a graphical modelling. The intention was not to return to a proposal that would remain speculative and that could not apply in the case of direct observation of nature: its role was first and foremost to allow easy identification of trees in the field, in the absence of flowers and leaves. The aim of the notion was to make it possible to identify the different "growth strategies"[7] affecting different species. By invoking the notion of "ideal shape", their aim – unlike that of Goethe – was no longer primarily that of trying to express a basic unit underlying all of living nature, i.e., a hypothetical common origin, by means of a single model but rather of trying to express and understand how, in the complex biological associations that occur in a tropical forest, the different "ideal shapes" emerge in a very stable and predictable manner, but then as a result of interactions on an ecological level subsequently drift away from their intrinsic ideals.

Although they were inflexible and rather formal, each "architectural model" was based on four characteristics that were entirely different from each other (for which reason it was initially impossible to quantify them) but botanically significant, which could be expressed by different "graphical symbols":[8] 1) type of growth (rhythmic or continuous); 2) branching structure (presence or absence of vegetative branching, sympodial or monopodial branching, or rhythmic, continuous or diffuse branching); 3) morphological differentiation of the axes (orthotropic or plagiotropic); and 4) position of sexuality (terminal or lateral flower). An "architectural model" in the sense intended by Hallé and Oldeman is therefore fully defined when a particular combination of these morphological characteristics and their related graphical symbols is obtained. Each combination found in nature (some are not) was given the name of a recognized botanist. These 24 combinations were determined by observation in the field and by monitoring meristem activity.[9] Hallé and Oldeman intentionally chose the concept of "model": it was looser than the concept of "type" as used in taxonomy. It was solely "based on the study of meristematic structures and functioning" and was therefore independent from the plants' biological type.[10] Yet it was precisely the relative "systematicness", the consideration of the overall shape from the point of view of its genesis and historicity, as well as the ability to reduce architectural variety to combinations of a small number of elementary rules, that made a convergence possible between this new concept of botany and the fragmented modelling method that de Reffye had used until then. This modelling linked series of choices and, in this respect at least, was suitable a priori for *computer simulation*, even if the model was not immediately suited to *algorithmic simulation*.

The search for botanical realism (1978–1979)

In 1978, when de Reffye, for his part, decided to prepare his thesis, all he had to do was reprise the method that he had previously developed, in an admittedly rather opportunistic and haphazard way, in the agronomic context that he was already specialized in: 1) fragmented modelling; 2) spatial simulation; 3) random simulation. Within a short space of time de Reffye realized that he would be able to reach the goal he had set himself: that of carrying out the most realistic simulation possible from a visual and botanical point of view, at least as far as the coffee plant was concerned.

To start with, he proposed a mathematical expression that made it possible to trace the statistical field measurements back to the stochastic processes that gave rise to the plant's first internodes. To this end, the material chosen at the outset consisted of young cuttings that were not yet affected by meristematic mortality. This made it possible to formulate and calibrate the probability of the *activity* of bud growth, or in other words the probability that a meristem would or would not form an internode. Next, de Reffye's work involved making this probabilistic law of elementary meristem growth activity more complex, so that his model would be valid for more than just very young plants and cuttings. To do this, the non-stationary nature of the growth activity as a function of branching order had to be taken into account. It should be recalled here that it was precisely the fact that this source of variability was not taken into account in Hisao Honda's purely geometric models that induced Jack B. Fisher to ultimately abandon graphical computer simulation. In this regard, however, de Reffye once again turned to the school of operational research. He resolutely entrenched himself in the formalism of stochastic processes. He was then able to take into account this non-stationary nature, which he decided to interpret as a simple *depreciation* of the growth probability in function of the meristem order on the axis under consideration: the variability of the parameters of the law of probability was thus accounted for by the well-defined concept of "stochastic process".[11]

Significantly, for each fragmentary step, de Reffye carried out comparisons between the simulations of the events of this step and the corresponding mathematical sub-model. In this way, he carried out an *empirical validation* of the simulation by comparison with a field data sample that had been deemed to be simple from an *empirical point of view* (the cuttings), which can be called a capture or grasping of the empirical by the simulation with verification of the simulation's conformity with the reality in the field. But, at the same time, he also effected a *theoretical validation* of the simulation by comparing it with the condensing analytical formulae (verification of the mathematical consistency of the simulated results) in a case that was deemed to be simple *from a theoretical point of view* (i.e., with standardized and unlinked elementary phenomena).

In the next phase the architecture of the plant was defined in greater detail. De Reffye followed the process he had used earlier in order to take into account this new biological phenomenon, which explained on a meristematic level the observable phenomena of failure or natural self-pruning of certain branches in the architecture. Using stochastic processes, he defined a mortality and a viability for the meristems. Since the viability did not coincide with the activity, he had to carry out rather long algebraic calculations of probability combinations in order to be able to analytically express the expected size and variation of a given stem. At this point, de Reffye repeated his method of *theoretical validation* of the simulation, since he was already in possession of the analytical formulae.

Lastly, de Reffye provided a diagram of the sub-program simulating the growth of a meristem with a given activity and viability. The program always proceeded meristem by meristem. It was designed to first carry out a probabilistic test (using the Monte-Carlo method) of the viability of the meristem under consideration; if this test was positive, the program would then carry out a probabilistic test of growth activity. In this way, the two phenomena could be easily intertwined in the new sub-program without the mathematical complications of analytical formulae: their intertwining could be processed sequentially since the program simulated the historicity of the complex phenomenon at the level of the individual life of a meristem. The program iconically (in the sense of mimetically) simulated the passage of time step by step at this level, precisely where the *intertwining* of these biological phenomena could be processed as an elementary *succession*. De Reffye then presented the values simulated by this new sub-program in comparison with the values that had been calculated using the "theoretical" formulae so as to confirm the sub-program's ability to numerically simulate failure or self-pruning. Next, he *calibrated* this numerical simulation model on coffee-plant clones that had been particularly meticulously measured and monitored – which was not necessarily the case for all the plants. The number of measurements necessary, or rather the number of stems to be counted on the various plants of a single clone, could amount to close to 2000[12] at times. Using his measurement tables and his simulated values, de Reffye concluded that "'random coffee plants' are then obtained that have the same behaviour as the plant observed in the field".[13]

There was still, however, another biological phenomenon that had not been taken into account and that could affect a plant's architecture at any given point in its history. This was the potential *dormancy* that could affect axillary buds, i.e., the buds that give rise to the lateral axes. An axillary bud that is formed during the creation of an internode on the stem might only begin to function some time later, after a certain delay. De Reffye therefore introduced the notion of "probability of functioning". A dormancy test was therefore included in the diagram of the new sub-program relating specifically to axillary buds. This test was carried out before the test for activity. In this case, too, the simulated and the theoretical values for a population of branches could be compared. All the sub-programs corresponding to the sub-models were then integrated by computer.

It must be noted that, at that time, de Reffye began to display a certain wariness with regard to the results of integrated simulation. It was for this reason that – as far as possible – he was careful to carry out step-by-step theoretical validations in the progressive construction of his integrative model of simulation. This in-principle wariness is clear in one section of his thesis, where he revisited what had been his own initial approach. He expressed his conviction that there must always be an "elementary tree" with the distinctive feature of being entirely calculable by hand and therefore capable of serving as a control for the entire simulation. The following extract from his thesis allows us to better understand the distinction between what I initially called empirical validation and theoretical validation.

> Verification of the correct functioning of the architecture simulation programs requires knowing how to theoretically solve at least one particular tree. This tree will be the "elementary tree". Although it has been relegated to a sub-chapter, the elementary tree is nonetheless at the root of the present work. Only once it has been thoroughly understood and solved can we attempt the approach with real trees. The elementary tree, and all those that simply derive from it, possess the property that all their architectural characteristics are calculable [. . .] We can then easily note the correct functioning of the simulations by observing the convergence between the simulated characteristics and their theoretical values.[14]

What I called "theoretical validation" is therefore in fact related to the *verification* of the program: it involves determining whether the simulation program is actually doing what it was expected to do from a formal point of view. In the cases where it can be uniformly formulated and calculated, the model may be used to verify the simulation. In fact, it is essentially the quality and accuracy of the computer implementation of the simulation model that are being tested,[15] whereas in the case of "empirical validation", it is an actual *validation*, i.e., a comparison between what the model gives and observable reality. De Reffye did not draw this distinction as clearly as subsequent computer engineers who specialized in modelling would later do.[16] But he was aware of the importance of program verification by pushing the analytical calculations as far as possible, in parallel with the simulations.

The last stage of the 1979 work involved first making the computer complete the 2D architecture outlines by adding fairly rough alphanumeric symbols to represent the leaves (two brackets side by side) and the fruit (two superposed asterisks).[17] In the next step, the transition to complete and actual simulation (in 3D) required the addition of two modular elements or sub-models to handle the geometric and mechanical aspects of the tree. With regard to the geometry, it was necessary, in particular, to factor in the branches' actual real-life phyllotaxis, which might be spiral or planar. De Reffye noted that any problem of this order could be dealt with as the rotation of a vector in space around the axis bearing the branch. He therefore explicitly adopted Karl Schimper and Alexander Braun's old notion of "angle of divergence" and wrote an analytic formula that would allow the new vector to be expressed as a function of the preceding one. It should be noted that this formula in itself did not pose any problem: it was enough for the formula to be recursive, since the processing of the axes in the simulation was exhaustive and stepped. The formula could even be made more complex, as required, at a later stage. In any case, it was easily supported by the software infrastructure. De Reffye could then use the computer to represent a tree bearing its own weight, from any viewing angle, and taking perspective into account.

By this stage he no longer spoke of modelling but rather of *reconstituting* the coffee plant. What did he mean by that? In effect, he viewed reconstitution as being essentially a visual replication. For de Reffye, complete simulation enabled a sort of qualitative validation, i.e., by eye, that would be valid on a global scale in this sense. It was for this reason that there were so many terms referring to vision in his work. According to him, the visual rendition of the first complete outlines already permitted a sort of validation of his work insofar as its biological value was concerned.[18] It should be recalled here that, traditionally, biology has always given considerable weight to observation,[19] and this was even more true in botany. De Reffye expected the botanists to agree. This interpretation is borne out by the following sentence: "It should be highlighted that the visual aspect of the coffee-plant is well rendered and that the IFCC experts recognize its validity".[20] De Reffye was appealing here to a sort of *argumentum ad verecundiam*. But the authorities he cited were, themselves, recognized, it would appear, and therefore his argument was sound, according to his reasoning. In fact, he was referring here to the hidden knowledge of experts. His program simulated the sense of the real that the expert has patiently acquired in the field: by replicating the real, he simulated the expert's competence or, in other words, the expertise itself. In order to do so, he considered it necessary to partially break down the traditional barrier between non-transferrable qualitative knowledge and transferrable knowledge that is merely quantitative. It should equally be noted that, here too, the model underlying the simulation can no longer be called mathematical. It was a mixed, pluriformalized model. As with previous work, it was the computer infrastructure that made the compatibility and interoperability between the mathematical sub-models possible in this case. In 1981 de Reffye would go further still and reveal the even more theoretically inspired underlying motivation (see the similar ideas of Eden or Cohen) that led him to consider that he was touching on something universal and deeply decisive. Indeed, he considered that the computer program was "no more than a translation of the [architectural] model's genetic program".[21]

Criticisms of theoretical models

As we know, the detailed criticisms of earlier plant architecture and growth models that de Reffye expressed at the start of his doctoral thesis cannot explain the driving force behind his achievement in his research work. It was clearly not these criticisms that led him to the computerized solution. We can therefore consider the criticisms to have been made essentially a posteriori. What must be clearly highlighted, however, is the fact that de Reffye belonged neither to the official theoretical biology clique that was forming in France at that time, driven by Pierre Delattre (1926–1985) in particular, nor to the circles of biological and ecological modellers from INRA or ORSTOM who, under the influence of luminaries such as Jean-Marie Legay (1925–2012), had in the meantime joined forces, in particular to deal with the concerted actions of DGRST.[22] But de Reffye's a posteriori criticisms are very useful for the case in hand, because – for the first time – they allow us to systematically put into perspective certain earlier, somewhat isolated, authors, whose practices and epistemologies I have occasionally highlighted, but without really seeing these authors converge with each other, and without any one of them approaching a method of modelling that was both universal and at the same time operational.

There is a common thread underlying all de Reffye's criticisms: the theoretical mathematical models that were produced during the first era of models could not have succeeded and they were only attractive because of an ignorance peculiar to the Western perception of plants. This was the ignorance in which we normally find ourselves with regard to the richness and diversity of plant architectures that are actually found in nature. A temperate-country inhabitant regularly comes across three architectural models at the most, whereas close to 24 can be found in a tropical forest. Usually, therefore, Westerners emphasize the branching process, whereas the meristem mortality processes are just as important in forming the plant architecture. De Reffye also ascribed his predecessors' interest in the analogy between biological trees and hydrographic networks (Horton, 1944; Leopold, 1971) to this over-emphasis on the role of branching and to the fact that they had assumed too quickly that one could generalize something that is generally observed only in temperate-climate trees: the morphological identity of the aerial plant axes.[23] This involved, once again, assuming too much internal homogeneity in trees. Horton and his successors had only tested their thermodynamics-based theoretical hypothesis on well-known trees such as the apple or cypress. By doing so, they believed that they would be able to establish a general rule for determining the number and length of branches on any given tree. In fact, contrary to Leopold's claims, they did not demonstrate, even roughly, the generality of their theoretical proposal. De Reffye suggested that any convergence between tree architectures in temperate areas and the architecture of hydrographic networks was therefore merely fortuitous. This convergence, if it actually existed, was not based on a true consideration of the optimal biological functioning of buds: "Hydrographical networks have random, dispersed branches, whereas tree branching can only come into effect from a localized bud".[24] Furthermore, there were architectural models, such as that of Roux (who in fact described the coffee plant architecture that de Reffye used as his starting point), where the axes could have a combined morphological identity: both orthotropic and plagiotropic at the same time. The trunk is orthotropic, but the branches are plagiotropic.

Having criticized the thermodynamic approach of "Horton's Law", de Reffye turned to the even older method of determining ramification angles and branch cross-sections by means of a physical and physiological theorization of the phenomena of vascularization in living beings. This approach was based on the principle that hydraulic flow energy, such as the one produced by the friction force in blood vessels, could be homogenized and thus dealt with in *one single optimization equation* along with the metabolic energy used by the organism in keeping the volume of blood in the vessels constant: this was "Murray's Law", from the name of the British biologist Cecil D. Murray (1897–1935) who developed the formula in 1926 when he was employed in Columbia University's Department of Physiology in New York. This type of formalization was also based, as we can see, on the hypothesis of a relative simplicity in formulating the *mathematical optimization functions* that are assumed to express *the optimality of the biological functions* involved in the morphogenesis of living beings. This law, which was later (from the 1980s onwards) called a mathematical "model", had certainly been fairly well verified experimentally. But it was based on a principle of optimization that in fact was rather questionable. The hypothesis of homogeneity and of the search for an optimum on which the law was based was in fact rapidly and vigorously contested by the physicist and engineer Paul S. Bauer (1904–1977), of Harvard's Fatigue Laboratory.[25] But the law had the advantage, in the case of blood-vessel radii for example, of arriving at a simple expression of the ratio between the three radii found in the presence of a bifurcation: $a_0^3 = a_1^3 + a_2^3$. For de Reffye, this method was still significant in 1979, since the law that it also arrived at for angles was well observed in the context of vascularization. Nonetheless, de Reffye noted that Murray had taken absolutely no account of the natural curve of branches in the case of arborescence in plants: the principle stating that branching angle minimizes mechanical work during the transport of sap would in any case be invalid if the branch curved *immediately after* its insertion either as a result of gravity or due to a genetic tendency to orthotropy. To highlight how well known this fact was in botany, de Reffye pointed out that Leonardo da Vinci was already aware of it in his day.[26] It would therefore be necessary to dissociate that which intrinsically (which, for de Reffye, meant genetically) determined the plant at a given angle of branching from that which mechanically determined the plant at a given curvature (and therefore to go beyond the pointless opposition between the mechanistic view of d'Arcy Thompson and Murray and the evolutionary genetics view). This was why, in this more general context, de Reffye accorded such importance to his computer subroutine for taking account of the mechanical issues of plant lodging or bending under self-weight loading and breakage. While this subroutine had been conceived separately, it nonetheless worked in a closely intertwined manner with the other sub-models. Furthermore, botanists distinguish clearly between immediate branch growth (sylleptic branching) and delayed growth (proleptic branching). Since Murray's hypothesis also involved a deviation of the principal axis compared with its initial direction at the level of the axillary junction, it is difficult to see how, in order to conform to this rule, a proleptic branch might *afterwards* modify the already-determined direction of the principal axis. Nor did it seem that the principles followed by botanical morphogenesis would be content with the layout suggested by the vascular analogy. A simple "principle of physiological optimality" such as Murray's, which was itself based on the model of the physical principles of optimality, did not take account of the effects of delay in plant branching.

In my view, what de Reffye discovered here was the fact that optimality, while it may actually exist in living phenomena, is *essentially unstructured* and *delocalized* in these phenomena, both from a strictly spatial point of view (which contemporary neo-mathematicism claimed to have dealt with already in the late 1960s[27]), and also from a *temporal* point of view during ontogenesis. *It was this simultaneously spatial and temporal lack of global structure that therefore first had to be*

carefully dealt with by means of multiple formalisms and associated simulations.
In the end, as for Horton's Law, Murray's Law was no more valid for mixed axes
than it was in the case of reiterations where there may be a significant angle formed
between shoot and the main axis. Rashevsky's "mathematical law" (drawn up in
1944 and subsequently termed "model" in the 1960s) presented the same failings to
de Reffye's mind. Rashevsky's model was also based on the hypothesis that metabolic constraints are decisive in the genesis of plant branching. The law consists of
allowing that the metabolic flux F, in a branch of radius r, is simply proportional to
that branch's cross-section: $F = K \cdot r^2$. In this way, Rashevsky formalized another
of Leonardo da Vinci's observations, which suggested that the total surface area of
the cross-sections of branches of order K is equal to that of branches of order $K + 1$.

According to de Reffye, all these mathematical modellings inspired by physical
analogies or by a reduction to the mechanical properties of metabolism failed precisely because they wrongly believed that they had more or less grasped the *essentials*
of the morphenogenetic driver of plant architecture, whereas in fact these modellings
turned out, a posteriori, to be merely an ad hoc superficial veneer on a plant phenomenon that was much more complex and, for that reason, was only partially known at
that time. But the problem lay precisely in the fact that, until then, no-one had been
aware of their ignorance. The illusion had remained unchallenged for so long because
of this second-degree ignorance. Although de Reffye was not himself a botanist, he
knew and had already modelled tropical plants – especially the coffee plant – and he
understood that it was only tropical botany or, in other words, the forward-looking
and descriptive science of plant reality in all its diversity, that could be expected to
offer the necessary rectifications. In this regard he was in full agreement with Francis
Hallé. From the point of view of botany, which he invoked in the introduction to his
thesis, all these *mono-formalized* works ultimately deserved the same criticism, since
they had always considered the plant as a "theoretical object",[28] i.e., as an object
whose details could, without great loss, be viewed out of context, homogenized or
broken down to formal self-similarities (fractals) in the various representations and
scenarios advanced to explain their genesis. What de Reffye challenged, it could be
said, was this hypothesis of immediate abstractability with regard to the plant object.
The aim of these "theoretical models",[29] undeniably laudable though it may have
been, was of course to "explain"[30] branching. But with the advances in the related
descriptive science, it was clear to de Reffye that these models that the theoreticians
might induce researchers in the field (agronomists, botanists) to accept – at least as
rough but inspiring analogies – had now been clearly disproven in their pretentions
to describe even the most general of architectural processes. De Reffye therefore
stressed that it was necessary to learn to better know things before seeking to explain
them. This lesson in empiricism may have seemed naïve and clichéd, if it hadn't
come at a time when the opportunities for theoretical and speculative research
had been greatly increased by the appearance and proliferation of new physical-
isms (based on energy, entropy, information theory, systemic, etc.) and of new and
increasingly absorbing mathematicisms (based on catastrophes, fractals, categories,
etc.). The "model method" itself, as we know now, served at times as a screen for this
type of practice where the humility of an avowed modellistic epistemology scarcely

conceals a speculative neo-mathematicism that is often exaggerated and pointless, since it disregards axiomatic diversity – even though this diversity now underpins and surrounds it.

Having criticized the physicalist and mathematical theoretical models in a targeted, and at the same time generalized way, de Reffye then evoked with greater interest what he called the "Lindenmayer school". According to de Reffye, this school "seeks to understand the morphogenesis of biological beings in a general manner, based on an internal logic or developmental languages".[31] Whereas a purely mathematical model provides a description, and a physicalist theoretical model promises an "explanation" – i.e., an unfolding or deployment of the phenomenon's advance in its physical process; an elucidation of how it takes place – the logical theoretical model, on the contrary, offers an "understanding", in other words it displays and unravels the reasons behind the phenomenon's choices, decisions and "logical tests",[32] since these reasons can be formalized directly and uniquely into linguistic rules, i.e., rules that are logical or "cybernetic", to use de Reffye's own term. In this third type of modelling, the physical properties of the biological substrate are therefore disregarded in order to concentrate solely on the resulting logical properties. It is therefore indeed a theoretical modelling that must be interpreted in a non-physicalist sense, rather as a reflection of a series of decisions made by the human mind. Nevertheless, in these more or less anthropomorphic-inspired logical modellings, we are still dealing, as always, with theory. This means that the starting point is not the observed phenomenon, but rather a "logical model" that is prepared beforehand and only later compared with "biological realizations that are meant to function in accordance with analogous processes".[33] Whether this cybernetic and logical approach was aimed at understanding or at explanation by unintentional means, it always, as a matter of principle, looked at a general case that could be conceptualized from the outset, and for this reason it abstracted. Lastly, what de Reffye saw as being truly innovative in the Lindenmayer school's approach was in fact a quality that, as I showed earlier, Lindenmayer himself at first considered to be secondary; the ability of the formalism to lend itself easily to realistic graphical representations on the computer.[34] In this way, the comparison with reality was facilitated. It also had the advantage of being able to take account of the variable delays in branch growth. The approach was therefore more flexible than the physicalist modelling, which was unquestioningly causal and reductively mono-causal. As a modelling of the establishment of morphology through morphogenesis, it was a dynamic modelling that was also able to account temporally for a sequence of differentiated events, and this sequence of modelling was also temporal in reality. This *algorithmic simulation* partially supported what I call *resemblance of dynamic*.[35] This is why it can be said that these logical models simulate: they account historically for the actual historicity of the phenomenon. There was, however, a major problem. The first systems proposed by the Lindenmayer school were "certain": because they had been forced to base these systems on a rigorous recursive axiomatic system, they modelled the branching processes as if they were deterministic – which was clearly not the case, in light of the field observations, as far as organogenesis was concerned. It was therefore impossible to easily account for the phenotypic variability of the architecture. On the other hand, de Reffye pointed out the particular pertinence of the notion of "statistical phenotype"[36] in the view of botanists such as Hallé and Oldeman. Yet it was precisely this variability that de Reffye wished urgently to account for, firstly to further refine the harvest predictions, and then (for his doctoral thesis) in order to strive for even greater botanical realism.

Criticisms of biometric models

In the end, since he was aware of the complexity of field data, de Reffye felt closer to the mindset of the biometricians. Nonetheless, he did not directly align himself with their approach either. Admittedly, the undeniable advantage of biometric and statistic modelling is that it no longer treats the plant as a "theoretical object". Instead, it treats the plant as an object that is "distinct and studied on the level of its own morphology".[37] In keeping with the richness of field observations, biometric and statistic modelling allows the variability of the measured material to be retained. De Reffye therefore agreed with Legay's 1971 article on mistletoe architecture. In this article, Legay discussed the value of Rashevsky's sap-flow model.[38] Legay sought to generalize this model by applying it to all the branching axes of orders 1 to *n*. In order to validate the model, he used numerous experimental measurements, including the length, diameter and division number of the mistletoe branches. He then carried out a biometric-type approach. In closely studying the resulting tables, it appeared to him that it was the volume of the branches that remained approximately constant, rather than their cross-sections. Legay then concluded his article with a general discussion on the lessons to be drawn from this type of study. He contextualized the model by describing it as being clearly *instrumentalistic*. Regardless of their type, *the models were merely tools*; in this case, they were purpose-built instruments. They were aimed a priori at the detection and expression of a single aspect of reality: in our case, either a metabolic interpretation on an organic level (Rashevsky's law), or a physiological and cellular interpretation of plant morphogenesis (Legay's model). The goal of developing a generalized branching model therefore would have to be abandoned.

It is instructive, however, to highlight de Reffye's interpretation of this specific work. His reading showed a significant, very revealing, distortion with regard to the epistemological lesson we are meant to draw from it. De Reffye first praised Legay's work for its empirical and inductive approach aimed at recording a distinct architecture on the plant itself. Furthermore, mistletoe has a rather rare architectural form that required the development of an approach without reductive preconceptions. De Reffye admitted that the plant species under study had the virtue of steering the modeller towards one approach rather than another. In this way, he recognized the fact that the coffee plant had played the same role in his own case. Thereafter, however, de Reffye did not adhere to the idea that models were merely "purpose-built instruments". His interpretation of the results shown in the article was completely different from that of its author. Regarding the author's conclusion, de Reffye agreed only with his criticism of the *proposed models* and not with his criticism of the *notion of models in general*. Like Legay, de Reffye admitted unreservedly that the models inspired by Rashevsky were inadequate for describing the complexity of reality. But it was *Rashevsky's overly simplified model* that de Reffye considered at fault, whereas according to Legay it was *the whole mathematical model approach* that should at any rate be questioned, since its effect was always to simplify. Faced with the same results, their conclusions were thus diametrically opposed. De Reffye drew an epistemological conclusion from the article that was the opposite of Legay's. In view of the history of ideas, it is of course necessary to consider this contradiction cautiously: the contradiction was not the result of a precocious and programmatic lucidity on the part of de Reffye that was entirely uncommon for the times, since de Reffye was already in possession of a fragmented and calibrated universal model when he wrote his introduction. But, despite the fact that the introduction was written afterwards, it has the merit – because of its conscious and in-depth nature – of clearly revealing the nature of the theoretical and methodological debates on the role of growth models. It was because de Reffye had developed his simulations *in the field*, alongside agronomic experiments – and thus in direct competition with them – that he was constantly concerned with replication. Unlike Legay, he therefore relegated even partial physiological or biological explanation to second place. De Reffye's simulations acquired meaning from a perspective that was primarily operational rather than immediately cognitive (by contrast, we can see clearly here how the instrumentalist epistemology of the model-as-tool can sometimes become a formidable driver of conceptual prematurities, despite his denials on this matter): his use of the computer was closely linked to the development of *simulations in the field* that first appeared as an addition to the empirical field inputs. As a result of his earlier epistemological decision, de Reffye saw statistical distributions as a way to trace back to the probabilistic laws that conditioned them, and then to simulate them by synthesizing the data. It was because he was aiming above all to develop a rapidly effective tool for the selection of plants, in all their uniqueness and in the field, that de Reffye directed his mathematical model towards the simulation of growth and not towards a condensing explanation on the physiological level of the processes involved.[39]

From a botanical point of view, the main success of de Reffye's work on universal simulation lies in its ability to simulate the entirety of de Hallé's and Oldeman's various architectural models. With this approach centred on the individual bud (also known as a *bottom-up* approach), de Reffye demonstrated that it was possible, using just one program, to reconstruct all of the differentiated growth strategies present in the 24 architectural models. From an agronomic point of view, the success of this universal simulation lay not only in its ability to highlight and to enable the extraction of early characteristics of the fruit-bearing capacities of the clones, as well as the associated inter- and intra-clonal variabilities, but also in its ability to precisely predict the average annual yield of any given coffee-plant clone: de Reffye found a correlation of 98 per cent between what was observed and what was simulated. Nonetheless, although this modelling and simulation work demonstrated that architecture is well-defined at a given age, it also demonstrated that, for any given individual, flowering – and therefore yield – remained heavily dependent on climate. This could account for some of the anomalies in the agronomic field testing. Optimization based on the relationship between architecture and yield remained possible if it was conceived on the level of mathematical expected values. Therefore, even though de Reffye's modelling solution had shown itself to be capable of explaining and accurately avoiding the shortfalls of Fisher's "design of experiments" method, it seemed unlikely that his solution would immediately and unconditionally replace it. It was therefore necessary to further improve how interactions with the environment during growth were taken into account. In addition, and related to this, was the fact that the simulation of tree stands (since the plants do not grow in isolation), rather than just the simulation of individuals, turned out to be beyond the abilities of the HP 9825, even when provided with memory upgrades. This was one of the first failures de Reffye had encountered: his simulation method, while it was calibrated in the field and became very botanically realistic, did not appear to allow a transition to the scale of stands or of plantations or, in other words, to a truly agronomic scale.

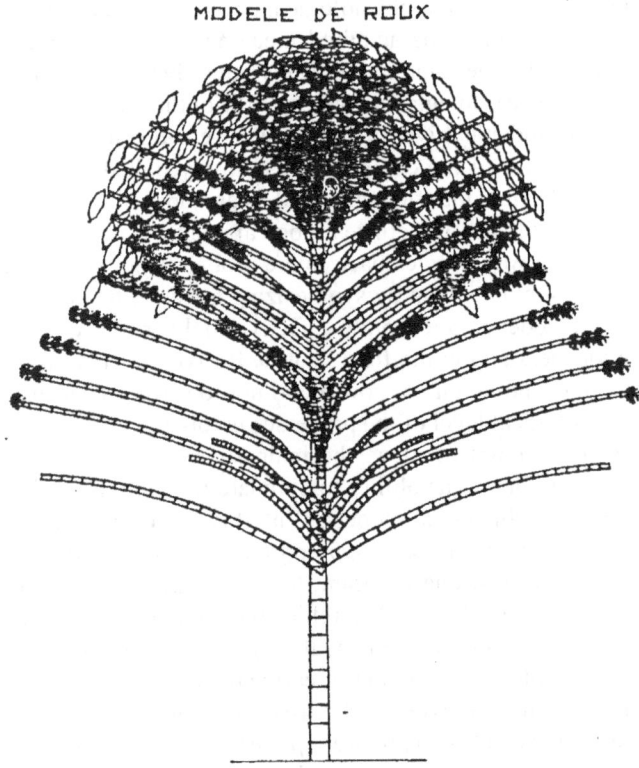

MODELE DE ROUX

Figure 4.1 Coffee plant drawn by plotter (Roux's architectural model). The leaf nodes and fruit-bearing nodes can be clearly distinguished. The curve of the axes is also calculated realistically. Taken from P. de Reffye's thesis, 1979, p. 166. Reproduced by kind permission of the author.

A mixed reception (1979–1981)

This program for the universal simulation of growth architecture rapidly achieved a certain success with botanists, despite their lack of familiarity with mathematical equations. In fact, the graphical simulations went a long way in persuading some of de Reffye's colleagues who were still very resistant to any formalization in botany: de Reffye thus discovered the formidable powers of communication and persuasion that these fragmented modelling images endowed him with. The confusion and admiration that these images aroused soon also spread to de Reffye's hierarchical superiors. They began to consider that, in de Reffye, they had a first-class researcher – even though in the two years that followed they continued to only employ him in the same context of applied science – despite the fact that de Reffye's approach had still not won unanimous support, since it inevitably met with criticism from the eco-physiologists. The latter criticized him for making a spectacle of science, and for indulging in stolidly descriptive modelling when he ought to have been concerning himself with the true mechanisms and factors involved in growth: temperature, humidity, nutrition, the mechanisms of bud functioning, etc. They were not at all convinced by de Reffye when he argued that his probabilistic modelling indirectly accounted for the genetic "causes", and that to this extent it would be superior to the phenomenological approach of classical statistical models.

At this time, de Reffye also experienced a failure that deeply affected him: he failed his INRA entry exam for the position of Research Fellow. He had decided to bank on his earlier work on stochastic and fragmented modelling of cocoa-tree pollination by insects. The fact that this work's application in agronomy was already indisputable was an important factor. He therefore drew up a summary for the selection panel: as was his custom, he presented a work on modelling that was split in two parts, one dealing with the moments favourable to the plant, and the other with insect behaviour, with the *computer combination of both modellings* making it possible to project pollination and therefore predict yield. But the selection panel, which was composed mostly of biologists and agronomists, seemed completely bewildered by this unusual approach: stochastic modelling using the queuing theory in particular was very largely unknown to them. The panel considered the document to be poorly presented and somewhat abstruse. The majority of the panel was therefore unconvinced. In fact, de Reffye was supported only by two researchers from the "plant sector"; Jean Bouchon and Alain Coléno, who at that time was Director of the "Department of Plant Pathology" at INRA. Coléno, a biometrician and statistician, recognized in particular the undeniable worth of the candidate's overall reasoning, while admitting that he wasn't able to grasp the particulars of the computer approach.

This episode illustrates fairly well how fragmented modelling, combined with visually realistic simulation by computer, often had a mixed reception at that time: the scientific value was not very clear to INRA. De Reffye remained disheartened for a long time by the selection panel's decision. At the time he took INRA's refusal deeply to heart. He even felt a certain bitterness regarding his research career, which he often questioned. Soon, however, he was offered an opportunity to return to France while remaining with his host institution, but as Director of Computing Resources. He decided to accept this offer, which risked distancing him indefinitely from research. Despite this rather unpromising or even alarming beginning from the point of view of the development of computer simulation, this stage of the history of plant modelling would nonetheless close on a remarkable recognition and considerable spread of its applications in numerous research sectors. After its emergence in the field of agronomy, simulation would become in its own right a *veritable field of experimentation* for many other domains. With de Reffye's return to France and the creation of his own laboratory, we can therefore say with good reason that there was a move from "simulation in the field" to the "field of simulation". Simulation would in effect become a field in two senses: it would be a modelling technique that would become the field, the site of completely new and favourable unions and convergences between approaches to plants that until then had been far removed from each other, such as computer graphics approaches, landscaping approaches, or the approaches of botany, forestry, silviculture, eco-physiology or arboriculture. During the 1990s many of these initially heterogeneous practices would one after the other feel compelled to opportunely take advantage of de Reffye's approach: these practices would join each other in that common place – that "common sense", in the less usual sense of the term – that new *common ground* represented by the realistic simulation of individual plants. This technical and institutional convergence became possible because in the meanwhile simulation had become, increasingly profoundly, a ground in another sense: it was *the ground of virtual experimentation*. Simulation was a duplicate of a complex empirical reality, which it could sometimes advantageously replace for these various disciplines or practices. At roughly the same time, these practices also became aware that they must consider the plant on an individual level and that, in order to collaborate, they could clearly not make do with sharing a common "Creole language" but instead must share even their *common substitute object*. No doubt, this notion of *virtual experimentation* may have seemed shocking and excessive. Nonetheless, the authors of this convergence and consolidation of plant simulation in France would demand it – despite a considerable epistemological reticence that was often expressed in the philosophical epistemologies themselves – and would give it a very particular form of credibility that will have to be explored in its moment of creation. It will therefore be highly enlightening to chart the events and the institutional, conceptual and technical conditions that allowed the, on the face of it, rather unexpected volte-face, *decisive convergence* and consolidation: only they can explain how computer simulation would once and for all leave behind the phase of theoretical speculation by taking the place of older models, especially in France during the 1990s – despite the resistance (including a deep-rooted *epistemological iconoclasm*) that, in this new form, it would still encounter.

Notes

1   At the end of the 1920s Plantefol (1891–1983) had been openly hostile to the mathematicized approach to phyllotaxis. He had proposed an important and influential theory of leaf spirals that made it possible to avoid the use of mathematical formulae.

2   In Hallé's own words (private correspondence of 3 July 2004).

3   *Essay on the architecture and growth dynamics of tropical trees.*

4 Hallé (F.), Oldeman (R.A.A.), *Essai sur l'architecture et la dynamique de croissance des arbres tropicaux* [Essay on the architecture and dynamic growth of tropical trees], Paris, Masson, 1970, p. 3. Emphasis added by author.
5 Ibid., p. 4.
6 Hallé (F.), "Modèles architecturaux chez les arbres tropicaux [Architectural models for tropical trees"], in P. Delattre, M. Thellier (Eds), *Élaboration et justification des modèles* [Devising and justifying models], Paris, Maloine, 1979, p. 546.
7 The use of the idea of "growth strategy" to designate the architectural model comes from a student of Hallé, the botanist Claude Edelin. This idea can be seen in the thesis he presented at Montpellier in 1977: "Images of conifer architecture". See Reffye (de) (P.), "Modélisation de l'architecture des arbres par des processus stochastiques. Simulation spatiale des modèles tropicaux sous l'effet de la pesanteur. Application au Coffea Robusta" [Modelling the architecture of trees by stochastic processes. Spatial simulation of tropical models under the effect of weight. Application in *Coffea Robusta*], Doctoral Thesis, Paris-Sud University, Orsay Faculty, 1979, p. 7.
8 In the words of F. Hallé, 1979, art. cit., p. 538.
9 By "recording the successive stages of morphogenesis in the form of sketches and photographic documentation", Hallé (F.), Oldeman (R.A.A.), 1970, op. cit., p. 5.
10 Ibid., p. 9–10.
11 Reffye (de) (P.), 1979, op. cit., p. 18.
12 Ibid., p. 71.
13 Ibid., p. 74.
14 Ibid., p. 114.
15 But other techniques existed and would be further developed during the period from 1980 to 1990, such as measurement of the statistical qualities of simulations, for example.
16 See, for example, Coquillard (P.) and Hill (D.R.C.), *Modélisations et simulations d'écosystèmes* [Models and simulations of ecosystems], Paris, Masson, 1997.
17 It is here that the glaring difference can be seen between the simulations of Cohen (1967) and Honda (1971), and those of de Reffye (1979). Cohen and Honda benefited from having the best graphic display screens of the time. In 1979, however, de Reffye was still far from being able to compete in terms of graphical rendering, and yet his simulations provided much more accurate botanical representations.
18 This idea can be equated to the idea of "empirical realism" expressed by Sergio Sismondo. See his "Simulation as a new style of research: iteration, integration, and instability", in G. Gramelsberger, Ed., *From Science to Computational Science*, Zürich, Diaphanes, 2011, pp. 151–163 and, in particular, p. 157: "what I am calling 'empirical realism' in the sciences depicts the world as it is found, rather than deep structures, hidden features, or created phenomena".
19 For this observation, I concur with E.F. Keller in *Making Sense of Life. Explaining Biological Development with Models, Metaphors and Machines*, Cambridge, Harvard University Press, 2002, 2nd Edition: 2003, Chapter 7.
20 Reffye (de) (P.), 1979, op. cit., p. 111.
21 Reffye (de) (P.), "Modèle mathématique aléatoire et simulation de la croissance et de l'architecture du caféier *Robusta*. 1ère partie. Étude du fonctionnement des méristèmes et de la croissance des axes végétatifs" ["Random mathematical model and simulation of the growth and architecture of the *Robusta* coffee plant. Part 1. Study of the functioning of meristems and of the growth of the vegetative axes"], *Café, Cacao, Thé*, 1981, Vol. 25, No. 2, p. 83.
22 *Délégation générale à la recherche scientifique et technique* (General delegation for scientific and technical research).
23 Reffye (de) (P.), 1979, op. cit., p. 7.
24 Ibid., p. 8.
25 In his short article of 1930, dedicated entirely to countering Murray's arguments, Bauer deemed Murray's use of a "principle of minimal work" a fallacy, because Murray's

example did not involve a conservative physical system (non-open from an energy point of view) and, as a result, *no principle of minimum work* could be applied a priori in that form to a biological-type morphogenetic phenomenon. See Bauer (P.S.), "The validity of minimal principles in physiology", *The Journal of General Physiology*, 1930, Vol. 13, July, p. 617.

26 Vinci (da) (L.), *The Notebooks of Leonardo da Vinci*, English translation by Edward MacCurdy, Reynal and Hitchcock, New York, 1939; Reprint: Braziller, New York, 1955, p. 306: "The branches of plants form a curve at every commencement of a tiny branch, and as this other branch is produced they bifurcate".

27 Such as the mathematicisms inspired by the catastrophe theory or a generalized information theory.

28 Reffye (de) (P.), 1979, p. 9.

29 Ibid.

30 Ibid.

31 Ibid.

32 Ibid., p. 10.

33 Ibid.

34 "The graphical implication of this formula [by Frijters and Lindenmayer] is produced by a computer. They obtain shapes that are fairly evocative of branchings", Ibid., p. 9.

35 By this term, I wish to highlight two correlated but differing properties that occur in such a simulation: 1) the segments of time in the target system are *iconically represented* (referred to) by segments of time in the computation of the simulation model; and 2) these segments of time in the computer run are *ordered in the same way* as the segments of time they are representing in the system. Hence the iconicity manifests itself on two distinct levels (material or substantive on the one hand, and structural on the other), which is not at all necessary for any given computer simulation.

36 Ibid.

37 Ibid.

38 Legay (J.M.), "Contribution à l'étude de la forme des plantes: discussion d'un modèle de ramification" [Contribution to the study of plant shape: discussion of a branching model], *Bulletin of Mathematical Biophysics*, 1971, Vol. 33, No. 3, p. 387.

39 Reffye (de) (P.), 1979, op. cit., p. 15: "The aim of the present work is to approach architecture on a numerical level by taking the greatest possible account of botanical data regarding bud distribution and functioning in a given plant".

5 Convergence between integrative simulation and computer graphics

In 1985, after an almost five-year break while development lay dormant, an encounter with computer graphics, in the person of Jean Françon from the Louis Pasteur University in Strasbourg, helped to spectacularly kick-start the research into universal architectural simulation at what was now known as CIRAD.[1] This convergence may ultimately have been somewhat premature, but, besides the software products that soon stemmed from it, CIRAD also quickly realized the various advantages to be gained in keeping and enhancing this less directly applicable research within their organization. As we shall see, however, the determination to make simulation conform to botanical reality never wavered; indeed, it was this demand for physiological realism that, in turn, stimulated further developments in computing and botany (e.g., software with parallel processing of buds). In this context, the issue of how to validate these simulations that were becoming considerably more complex therefore became a major technological and epistemological concern.

In 1980 the decision was made to create GERDAT, the *Groupement d'étude et de recherche pour le développement de l'agronomie tropicale*.[2] Henry-Hervé Bichat, an agricultural engineer, was named Director of this new "EPIC" (*Établissement Public à Caractère Industriel et Commercial*[3]), which combined the ten or so tropical research institutes, including IFCC (which was renamed IRCC – *Institut de Recherche sur le Café, le Cacao et autres plantes stimulantes*[4] – in 1983), that were dedicated to various branches of production: plant, animal, forestry, coffee, etc. Bichat worked for three years on transforming GERDAT into what, in 1984, became CIRAD (*Centre de Coopération Internationale en Recherche Agronomique pour le Développement*[5]). He soon saw the necessity for and advantage of computerizing his institution at the earliest possible opportunity. De Reffye's competence in this field was put to good use: he was named Director of Computing Resources at GERDAT and then at CIRAD. This meant that his research work in plant modelling was brought to a halt for a time, while at the same time ensuring him an important position within his institution.

During this pause in his research de Reffye nonetheless pushed himself to publish his thesis, which he did in four successive issues of the *Café, Cacao, Thé* journal between June 1981 and March 1983. But, just as with the earlier articles on modelling that he had published in the same technical journal, the

DOI: 10.4324/9781315159904-6

initial response was rather poor despite the journal's articles being listed in the *Current Contents* international indexing system. But de Reffye's articles were in French, and therefore remained somewhat inaccessible to foreign researchers. In his spare time, with the help of computer scientist Joël Sor, who de Reffye had engaged at CIRAD, he translated his 1979 program into FORTRAN so as to be able to use it on other machines.

The relaunch of research into architectural simulation (1985–1991)

This was how things stood in the autumn of 1985, when a computer scientist from the Louis Pasteur University in Strasbourg, Jean Françon (born 1936), contacted de Reffye directly by telephone. This decisive conversation compelled de Reffye to return to full-time research. Let us return briefly to Françon's career and his own motivations. Following his university studies in mathematics and a Master's degree in Physics in the early 1960s, Françon had become an engineer in the Strasbourg computing centre (linked to the Strasbourg Nuclear Research Centre, which later became part of the IN2P3[6]). In 1979 he had presented his doctoral thesis in enumerative combinatorics, focusing in particular on algorithmic problems linked to questions of sorting. To Françon's mind, at that time, computer science was an entirely separate science that, as such, required mathematics but should not be confused with it. For him, computer science had a status comparable with that of the natural sciences, such as physics, in its relationship with mathematics. At the time, Françon's views on this matter were very unusual in France. Nevertheless, in 1980 he was appointed Professor of Computer Science at the University of Haute Alsace in Mulhouse, and then in 1985 at the Louis Pasteur University in Strasbourg, where he primarily taught computer graphics. In an effort to connect the thinking on theoretical computer science with the solutions that existed in nature or in the theories of natural science, Françon closely studied Horton, Leopold, Strahler and the combinatorialists who developed the Strahler number theory – a branching theory that he considered to be closely related to sorting issues in computing. The "algorithmic" therefore had to learn lessons from the school of natural science: the fact that it had already done so for certain problems regarding sorting or data search was proof enough for Françon that algorithms were not reducible to a subsector of mathematics.

Between 1983 and 1984, Françon heard of the various works in computer graphics that were starting to be developed based on Lindenmayer systems. He suggested that enumerative combinatorics – the science of formal trees – might do an even better job than L-systems or than the simple geometric self-similarity relations of fractals.[7] He felt that he could contribute something new. During this period, he and his doctoral students threw themselves into work on what he called at that time the "combinatorial modelling" of figurative plants. Since there were various efficient enumerative combinatorics theorems or algorithms that enabled a binary tree of a given size to be extracted directly from the set of all binary trees of that size, it seemed that it might be advantageous to bypass the generative and step-by-step nature of L-systems on this point. In this way, we return, once more, from simulation to modelling. Françon steered one of his students, Georges Eyrolles, down this route. In 1986 Eyrolles presented his doctoral thesis on the "Synthesis of figurative images of trees using combinatorial methods".[8] This was clearly combinatorial modelling: it did not primarily seek exact conformity to botanical reality, but simply an approximate conformity. Once again, the stated aim remained the convenience and speed of the algorithms. In this context, the notion of "figurative image" referred to the computer-generated images that "figure" a plant, creating an illusion for the uninitiated, i.e., images that qualitatively and very broadly resembled a real plant but that still differed widely from images that faithfully rendered the botanical detail.

In May 1984, however, the *Computer Graphics and Applications* journal of the IEEE (Institute of Electrical and Electronics Engineers) published an important article by two Japanese authors, Masaki Aono and Tosiyasu L. Kunii,[9] an engineer with the IBM-Japan Institute of Scientific Research and a teacher-researcher in the Computer Department of the Tokyo Science Faculty, respectively. This in-depth work, with its impressive computer-generated images, summarized the a priori mathematical modelling approaches, whether these were of combinatoric, logical (Lindenmayer) or geometric (Honda and Fisher) type. The declared aim of Aono and Kunii was to produce very botanically realistic images of trees. Having demonstrated the excessive rigidity of L-systems, they instead used the formalism of binary and ternary trees by combining it with geometric rules (branching angles, elongation ratios) similar to those of Hisao Honda. Aono and Kunii ensured the very realistic nature of their geometric simulations by carrying out comparisons with photographs of real plants, and these comparisons were evaluated using statistical techniques, meaning that their geometric models were calibrated a posteriori. It was thus by using a *top-down* method, so to speak, that they made the mathematical models more flexible, whereas de Reffye, from the outset, had used an optimal *bottom-up* adaptation, i.e., on the basis of the meristems. At any rate, what this article demonstrated very clearly was that, even when the intention was to retain a theoretical modelling approach, the purely combinatorial solution already appeared to be out of date.

In fact, Françon's aim had always been to achieve botanical reality: he would have liked the backing of real botanists. Aono and Kunii, however, were still concentrating only on computer graphics. They used programmer "tricks", although obviously using techniques that were highly advanced and that, above all, were completely inaccessible to the French university from a financial point of view. From the point of view of scientific approach, however, they can at the very least be criticized for having ignored botanical reality. Nonetheless, at that point in time Françon could not see how to rid himself of the dissatisfaction he felt with regard to the a priori mathematical models that were being extensively used by his colleagues. In the end, it was his colleagues in botany who, in response to his insistent demands, introduced him to the only publications by de Reffye that were available at that time: those in the *Café, Cacao, Thé* journal.

It was a shock for Françon. To his mind, de Reffye was proposing something better than a model since he *replicated the details*. From that point on, almost in a spirit of provocation and even counter to the opinion of de Reffye himself, who considered himself far from having completed his task, Françon constantly touted the idea of de Reffye as the "Newton of the blade of grass".[10] In 1991, in his presentation at the Montpellier Symposium on *The Tree*, after a brief introduction to plant modelling using L-systems, Françon declared:

> [de Reffye's aim] requires that the modelling be validated in all its details by precise observations and measurements (not just metric measurements, but especially measurements of durations of elongation) in the field, which no other method has done. This is the methodology of physical science. This is why I consider that Philippe de Reffye's modelling should be ranked as an experimentally validated theory.[11]

In 1985, therefore, Françon impressed on de Reffye the idea that computer graphics was in a rut and that it needed his engineering approach to counter all the tricks used by programmers to "make it real", which Françon condemned as lacking in theoretical value. De Reffye responded that his visualization software had not been conceived specifically to compete with the solutions of computer graphics specialists, and that visual simulation served above all to *prove* to botanists and agronomists the validity of the underlying computer model. Nevertheless, de Reffye recognized a considerable convergence of interests in Françon's proposed alliance, even though the interests of each were different: for de Reffye, accepting this collaboration would be a way to continue updating and extending his initial software in order to see whether it could become more effective and therefore better accepted, particularly in agronomy. This decisive merger took place very rapidly, first with the arrival of two doctoral students in computer science, followed by a relaunch of the research into architectural simulation, and finally with the creation of a laboratory dedicated to this work within CIRAD.

Jaeger's thesis: the prefixed model and synthesis of botanical images (1987)

Marc Jaeger (born 1962) was the first of Françon's doctoral students in this field, but he principally worked on site at CIRAD Montpellier. Due to financial constraints, the Strasbourg university laboratory had become outdated. CIRAD, on the contrary, was already able in 1985 to supply Jaeger with a Data-General minicomputer and a Tektronix monitor. Jaeger's instructions were to use the algorithms from de Reffye's thesis, to program them in FORTRAN, and as quickly as possible to supply images that could be shown at an upcoming SIGGRAPH[12] symposium to compete with those already obtained, in particular by Aono and Kunii. It was necessary to demonstrate the value of this solution as soon as possible. However, having a publication accepted at SIGGRAPH was rather rare for the French: most of their proposals were rejected as competition was fierce. The aim for this first two-discipline thesis was therefore to prove the value of the solution and to publicize it. This was one of the reasons why the botanical theory was not further refined compared with de Reffye's original thesis. In order to formalize this research activity, CIRAD created a new "plant modelling laboratory" in 1985, as part of their "computing centre", which had then just been newly renamed "GERDAT".[13] Since Françon knew the habits and customs of the computing and computer-graphics world, he immediately advised CIRAD to give the laboratory a catchy name, which could also serve as a logo for the software they expected to produce. De Reffye proposed calling it AMAP: *Atelier de Modélisation de l'Architecture des Plantes* (Plant architecture modelling workshop). At the beginning, AMAP consisted only of de Reffye, Marc Jaeger, René Lecoustre – an agronomist who had worked on pollination with de Reffye and had returned at the same time as him from Côte d'Ivoire – and Evelyne Costes, a young botanist who was working on her doctorate on the architectural analysis and modelling of lychees at the University of Science and Technology of Languedoc (USTL) in Montpellier. At USTL, Costes worked with Francis Hallé, who had in the meantime become Director of the Botany Laboratory, and alongside the botanist Claude Edelin, who had just completed his two theses at the same university, one on conifer architecture (his first doctoral thesis in 1977) and the other on monopodial architecture and automatically repeating shapes (his second doctoral thesis in 1984). Having recently been recruited to USTL, Edelin could therefore also closely follow the work of AMAP. The role of these "Hallé-school" botanists was first to specify and consolidate the concepts of simulation and extend them to include new plants. AMAP thus came into being based on a collaboration between CIRAD, the Montpellier Institute of Botany, ULP, and the Laboratory of Computer Research (CNRS) at Paris-Sud, directed by Françon's colleague, Claude Puech. Puech would thus be part of the panel reviewing Jaeger's thesis.

In this first work, Jaeger used a constructive *bottom-up* method and, for that reason, procedural programming[14] was chosen rather than an object-oriented programming[15] based on hierarchically ordered classes of objects – a choice dictated by time constraints, although object-oriented programming was available at the time and may have offered a better solution. This procedural approach reworked the construction of the plant based on the stochastic modelling of the meristems, i.e., based on the detail and not on previously established definitions of intermediary-level botanical objects.

One of the major conceptual contributions of Jaeger's thesis lay in the clear and explicit distinction between the plant *topology*, which is controlled by the growth engine, and its *geometry*, which is controlled by an autonomous module that is compatible with a standard graphics visualization tool. In this way, he could further clarify – and adapt to real botanical data – a distinction that had been introduced previously by a *ComputerGraphics Lab* engineer, Alvy Ray Smith, who had used it merely to try to resolve by computer the old problem of the geometric interpretation of L-systems.[16] In his work, Jaeger therefore used high-level and structured languages – first FORTRAN, and later C – to begin slowly resolving what, until then, had been the difficult or impossible issue of communication between the various formalisms (strictly topological, probabilistic or geometric). Following in de Reffye's steps, Jaeger discovered that the translation of an algorithm into a differently structured language made it necessary to find different ways to conceive and construct the model by computer. The computer structures therefore served as the sites of formal mediation, i.e., as the scene where schemes of numerous micro-actions and micro-interactions were delegated to the computer, from whence they mediated between heterogeneous systems of axioms. The model was not just translated into a computer language: through programming, it also became clearer, more complex and communicating – indeed its very existence was made possible, in all the diversity of its conception itself. The act of programming here was thus not just a simple procedure of translation between two different languages (mathematic on the one hand, and computational on the other). Producing such a model using this type of programming structure was not merely a mediating transaction that operated mainly on a linguistic level, as – on the contrary – a simple *trading zone*, in the sense intended by Galison, could do. In the case of AMAP, what the structured computerization soon made clear was that Hallé and Oldeman's highly qualitative classification should perhaps in fact be reorganized, as Claude Edelin had suggested at the time.[17]

This first commercial version of the software did not yet offer resemblance of dynamic simulation. The simulation dynamics cannot be matched to the plant-growth dynamics at each time-step,[18] because this would require all the nodes to be processed in parallel, as is the case in reality, since the meristems of different branches of a real plant evolve in a simultaneous, correlated and non-sequential manner. Time passes "at the same time" for these meristems, one might say. The solution that Jaeger first came up with, however, was to generate the plant, node by node, at a given age (fixed beforehand). It was only once all the nodes on a given axis and its branches had died or reached a specific age limit that what the AMAP researchers called the mathematical and computing "growth engine" would move on to the neighbouring axis. Thus, despite the existence of de Reffye's by-then already old proposals on this issue, the solution of true recursiveness was not chosen. This solution was simple to manipulate and required very little memory resources. But the transitional shape obtained during calculation could not be displayed, as it did not resemble a real tree. It remained impossible to simulate any obstruction between axes since each axis was "calculated" right to the end before moving on to the next, and each meristem was therefore "unaware" of its environment. Thus the simulation was not yet botanically realistic in its dynamic *process* itself, even though it was fairly rigorously realistic in its various final *static results*, at each determined final age.

As a result of this work, by the mid-1980s the laboratory could offer researchers veritable "computer mock-ups",[19] as they were termed in a joint article by AMAP members, that could be viewed in three dimensions. These mock-ups were integrated into scenes that could include buildings, towns, etc. They could represent veritable virtual botanical gardens. For the first time, and in contrast to earlier attempts (such as that of Aono and Kunii), plant architecture simulations also had an actual botanical foundation. Some specialists, such as Hallé, put their trust in Jaeger, although they had remained sceptical of the purely theoretical and a priori tests based on fractals or L-systems. Yet Hallé was not resistant to computer approaches. He had even met Lindenmayer between 1975 and 1978, during some of his visits to his colleague Oldeman, who was employed in the Netherlands at the time. But the two researchers were not able to agree on the terms of a potential collaboration, partly because of Lindenmayer's overly formalistic approach. Jaeger and de Reffye, on the contrary, were more convincing to Hallé and his botany students because their simulations benefited from an even better visualization thanks to the new computer monitors. The principle of validation "by expert eye" that de Reffye had invoked was more valid than ever.

But, outside of Hallé's school, the reception among botanists remained mixed at the time. Agronomists and eco-physiologists were even less convinced. To be honest, the main interest of Jaeger's software resided above all in the fact that it could be used for applications that were not primarily agronomic, or even botanical, but instead were media-related. The software creators were deeply aware of this, as it would make promotion all the easier. Media interest in de Reffye and Jaeger's work was soon considerable. The software resulting from their thesis was sold under AMAP's name, with a CIRAD licence, and on the basis of contracts agreed with companies that specialized in computer-generated imaging, such as SESA.[20] It could thus be used to help in decision-making for landscape design professionals in urban planning. Certain computer-aided design (CAD) programs offered AMAP features for image synthesis. This work attracted considerable attention at the February 1987 *Forum des Nouvelles Images*[21] in Monte Carlo. Numerous images created using AMAP were published in mainstream magazines. The graphical results of the new approach played a large part in its popularity and especially in increasing the worldwide reputation of CIRAD. Agronomic concerns seemed far away.... Many Japanese clients used AMAP simulations in resolving problems in landscaping, botanical garden design or animation. Even though, as we saw, geometric simulation of plants was first developed in Japan, by the late 1980s the Japanese were still lacking in botanical accuracy. This was precisely what AMAP offered with its topological module integrating an approach by stochastic processes and AMAP therefore rapidly benefited from this enthusiasm. Thanks to its ability in promoting the results of its research, CIRAD made a profitable business transaction, which aroused some envy among competitors, however. Finally, in 1988, Puech and Françon succeeded in getting their first article on AMAP included in the SIGGRAPH[22] symposium. The year 1988, therefore, witnessed a sort of consecration of AMAP in the realm of computer graphics.

After this initial success in promotion, it became imperative to re-establish ties with the agronomists, and this only seemed possible by pursuing Françon and de Reffye's initial aim of making computer simulation increasingly realistic from a botanical point of view, in particular insofar as its process of generation was concerned: it was necessary to integrate the actual parallelism of bud functioning. At the time, this may have seemed the only route to reconciliation with the agronomists, since it would then be possible, by taking photosynthesis into account, to envisage showing, step by step, the physiological processes of biomass allocation. Since the mock-up was more realistic from the point of view of growth history, it could also be made more functional and less descriptive.

Figure 5.1 Simulation of a chestnut tree in winter. Philippe de Reffye, AMAP-CIRAD
software, 1992. Reproduced by kind permission of the author. For this edition
of the book, the 1992 software has been relaunched to compute this new
figure in order to have a better image but also to show what was possible in
those times.

Blaise's thesis: the simulation of bud parallelism (1991)

It was Jean Françon's second student, Frédéric Blaise, who between 1988 and 1991
was tasked with transition to the simulation of parallelism. A reorganization in the
CIRAD laboratory fostered study on an approach to simulation that could be used
by botanists and agronomists alike, irrespective of the image aesthetics. It ultimately
became clear with the prefixed approach (without trajectory matching) that there
was still not enough distinction in the meristem probabilities between what was due

to genotype and what was due to conditions in their surroundings. It was not certain, therefore, whether the values for these intrinsic probabilities (or those that were considered intrinsic) were good, even if the result was realistic. The local probabilities did not formalize an elementary biological phenomenon. The realism would have to become even more botanical instead of remaining merely superficial. The parts of the tree effectively blocked each other at every step of their growth, due either to mechanical blocking between neighbouring branches or to self-shading in the crown. Once this blocking was taken into consideration, the way became open not just for modelling epigenesis but also for modelling stands of plants (including plantations and forests) rather than just modelling individual plants. De Reffye thus rediscovered the aim he had held after completing his thesis.

This time, unlike Jaeger, Blaise chose to represent the entities in hierarchized structures so as to avoid informational redundancy in defining these entities; this was an important issue in a program that he expected to be rather cumbersome. He included six levels in his program: internode, growth unit, axis, reiteration, structure and plant. The entities were *structures* and their links were *pointers* in a C-language sense. It should be pointed out that, in programming terminology, pointers are the entity's memory locations that indicate the addresses of the following and preceding entities. They make it possible to dynamically organize allocations, classifications or storage. The lowest-level entity in this case was the internode and the highest was the plant. Compared with Jaeger's simulation, botanical precision had to be sacrificed to a certain extent so that the data structuring could be efficiently hierarchized. For reasons of computing feasibility (implementation in hierarchized language), some of the relationships between certain organs had to be systematized by sacrificing the fine details. Nonetheless, it was now possible to maintain a good level of realism thanks to a suggestion of introducing the notion of "growth unit",[23] which was already well established in botany. Indeed, following on from Claude Edelin's work, Evelyne Costes had in the meantime highlighted the pertinence of this notion for modelling in her thesis on lychees. She demonstrated that the first 1987 version of the software, which was still heavily influenced by coffee-plant growth, had confused the internode with the growth unit: she was not able to correctly simulate the lychee using this first program.

Along with this fragmentation into entities that now had a clear biological significance, Blaise chose to use object-oriented programming as well as an adapted simulation technique that had already been tried and tested, for that matter, in operational research in particular. This was the discrete event simulation technique. In this technique, the numerical variables describing the system are discrete and finite in number. There were three significant consequences as far as carrying out the programming was concerned: first, the set of combinations of these values formed the state space of the simulation and was, in principle, finite or, rather, countable. The second consequence was that the evolution of the system being modelled was itself discretized, since it was not possible to pass continuously from one discrete value to the next. There are therefore *instants of change*, called the "event occurrence time" or event dates, where the system is allowed to make a hop. As a result, time is also therefore discretized.[24] The third consequence of this discrete processing of the variables can be expressed as the notion of the processing order of simultaneous events. This order refers to the choice that was now open to the programmer to make events that took place simultaneously be processed in a sequential order. The order of events could thus be completely disregarded by the software user. This computerized discretization of time had the fundamental quality of enabling the programmer to "stop" time in order to process all the simultaneous tasks, one after the other. The simulation time was therefore a virtual time that was distinct from real time, not only because it had a different rhythmic relationship but also because we can choose to "stop" it in order to carry out parallel tasks. Using this technique, a sequential machine may thus function in an almost parallel fashion. Under these conditions, time management may be of two types: by clock or by event. In simulation time management by clock, at each instant of change from one time step to the next, the list of events is examined and all the events that may appear at that date are activated. Blaise rejected this solution, however, because the choice of elementary time steps was problematic as far as plants were concerned. A biological phenomenon is distinctly different in this matter from a mechanical or industrial phenomenon as regards how it may be conceived in advance or reconstructed by CAD.[25] If this phenomenon is not artificial, in other words if it is biological (contrary to what takes place in a mechanical engineer's or an architect's technical drawing, where it can be easily depicted as a module or elementary unit), there is no simple unit of time measurement for development since we were not given the opportunity to choose one. It is therefore not easy to assign a minimum event, i.e., a minimum duration that would constitute a basic time measurement for all the other events.[26] This is due to the fact that we do not have sufficient knowledge of a hypothetical minimum scale of natural phenomena to the point of being able to quantify them and reduce them to simple rules of arithmetic and logic.

Nonetheless, the notion of *internal clock* did not lose all pertinence, even though it was not actually implemented as such in the program. In effect, it was the notion of *schedule* that would identify the program functioning with the ticking of an internal clock. In a schedule, real clock time strokes can be included. It is *event-based simulation* that allows this. The simulation time is then managed by a "linear list of events", as Blaise[27] called it. When several events may take place at the same time, they call each other and follow each other in the schedule, without any change to their "date" field in the list of definitions of their attributes. Their order of apparition, which is invisible to the user, therefore depends on the way in which the schedule is scanned.

Following this new time management and the choices it imposed, the second generation of AMAP software also required a new consideration of the use of space. The recommended solution was to follow the trend to discretize nature. Space itself was therefore treated as a heap of elementary fixed-size cubes and the geometric coordinates of the plant were rounded-off in order to fit into one of these cubes: it was necessary to make the geometrical logical so that logic and geometry could become compatible at each step, but also so that the calculations could be carried out in a limited time. Indeed, analytical geometry solutions – although they may seem elegant and closer to the reality of elementary forms (internodes, leaves, branches, stems) – require a prohibitive use of calculations, given the number of elements that would have to be taken into consideration. For that reason, Blaise borrowed a concept of space known as a voxel from medical imaging techniques. A voxel is the three-dimensional analogue of the pixel. It is the smallest unit of volume visible on screen and processed as such by computer.[28] Essentially, the point of this discretization was to replace metrical relationships with logical relationships, such as presence/absence. These relationships were easier to manage for the algorithms dealing with branching and stochastic growth. It was therefore useful to discretize at the very beginning of the process and not just at the last phase of on-screen display in pixels. In this way, the parallelism of the processing (the unceasing transition from topological to metrical) was in turn facilitated. But it was also necessary, by means of these modelling conditions, to make sure that the performance of the computer tool remained available. In other words, it was a compromise solution.

To demonstrate that his choice of spatial discretization step did not create any artefacts compared with the model to be simulated, Blaise carried out reiterated simulation calculations in which only the step varied. By means of these reiterated *simulation experiments* and by noting the *stability* of the simulations obtained from a statistical point of view, Blaise was able to conclude that the discretization of space was correctly adapted to the objectives. It was in this sense that Blaise spoke of "experiments".[29] He tested the statistical properties of the simulations with a commonly used software for statistical analysis, SAS, as one would do for real experimental designs. This work was part of the *verification* stage (the harmlessness of the implementation with regard to the model).

How can an integrative simulation be validated?

In the end, the validation method that appeared to satisfy the AMAP team involved simulating a large number of different architectural models and displaying the resulting images. The demonstration thus relied on the realistic nature of certain plant images. In this instance too, Blaise allowed himself to speak of "experiments" when talking of this type of validation through the realism of the final result. But care must be taken to no longer use the word "realism" here only in the sense of a one-off resemblance of fixed images to a snapshot of reality. Instead, it is in fact the rendering as a whole of the plant's vigorousness, its growth, conflicts, adaptive strategies in the presence of obstacles, i.e., its history, that is evaluated in terms of realism. As we have seen, this software undertook to give a realistic growth dynamic, and not just snapshots. It was therefore the growth engine itself that could be evaluated, and not some unspecified image of a plant.

Finally, in accordance with the intended objective, this generation of software – known as AMAPpara (for "parallelism") – made it possible to simulate the secondary growth, i.e., growth in thickness, of the trunk and branches through the transport and deposit of assimilates. The software also allowed interactions between the trees and their surroundings to be taken into account.[30] Furthermore, AMAPpara was evolutive, thanks to the relatively complex structuring of its data compared with the earlier version. This complexity was a positive sign according to AMAP's botanical and agronomical researchers, since it was intended to reflect the complexity of the actual plant.[31] With realistic computer simulation, the complexity of the model therefore no longer necessarily signified falseness or uselessness. The minimalist epistemology of the model (whereby the model must be minimal, i.e., abstractive, in order to be useful) seemed to have been well and truly banished by this work.

There were nonetheless limitations to this simulation. These arose in part from the frequent approximations that it used. Above all, however, they arose from the fact that pure parallelism had not yet been entirely achieved in simulation. Thus, the fact that a schedule must always be scanned in the same direction and at the same time carried a risk of introducing into the architecture certain regularities that would become simulation artefacts.[32] In order to counter this unfortunate consequence, it was necessary to introduce some artificial randomness in the path of the schedule. Finally, it should be noted that the discretization of space into voxels entailed a search for a compromise between excessive memory use and the level of detail of the step. Be that as it may, Frédéric Blaise submitted his thesis in 1991 and was subsequently recruited as Research Fellow at AMAP.

During the preparation of Blaise's thesis, de Reffye had been spurred to further improve the distinctions between botanical concepts. At this time, he proposed reusing an old concept, first advanced in 1965 by his ex-professor at ENSAT, Pierre Rivals (1911–1979): the physiological age of the meristem. Put simply, the probabilities of different meristem activities depend in effect on the meristem's location (i.e., its order) in the plant topology: the higher the meristem's order, the lower its vigour. It can therefore be considered that it comes into being with a certain age – its physiological age. De Reffye gradually returned to this idea because he had become aware of the reflections of a young research engineer with a doctorate in statistics who had just been employed by AMAP: Eric Elguero. Elguero[33] immediately recognized a particular type of point process behind meristem functioning: a Poisson process that could be expressed as a renewal process.[34] Since the comparison with this type of process made it possible to combine the different Poisson processes into one single expression, the notion of physiological age was therefore reintroduced in 1991. This botanical event came about shortly after the notion of "reference axis" had also been proposed by de Reffye and the botanist Daniel Barthélémy, based on an architectural simulation of the Japanese elm. In the meantime, Barthélémy had presented his botany thesis[35] in 1988 at USTL, under the supervision of Francis Hallé. He had proposed the idea of "automatic flowering" in order to explain the sexuality of a number of tropical plants.[36] Partly inspired by the earlier idea of automatic flowering, the "reference axis" evolved as a theoretical construct "based on the grouping and classifying of all the stages of differentiation of a tree".[37] These differentiation stages, in turn, were based on the notion of physiological age and the new formalism linked to it. The reference axis in fact translates the progression of this age along a theoretical axis. It is an automaton that translates the change of functioning of the meristems. Because of its automatic, recursive and general nature, the automaton was therefore able to help considerably simplify the computational growth engine by giving the impression of a great cohesiveness (of a computing type) beyond the diversity of the architectural models that, until then, had tended to result from the intertwining of locally simulated probabilistic laws.

Notes

1 *Centre de coopération internationale en recherche agronomique pour le développement* [French Agricultural Research Centre for International Development].
2 Study and Research Group for Tropical Agronomy.
3 Public-Sector Industrial and Commercial Enterprise – a type of public body established by statute in France.
4 Institute for Research on Coffee, Cocoa and other Stimulating Plants.
5 Agricultural Research Centre for International Development.
6 The *Institut national de physique nucléaire et de physique des particules* [National Institute of Nuclear Physics and Particle Physics].
7 See Françon (J.), "Sur la modélisation informatique de l'architecture et du développement des végétaux" [On computer modelling of plant architecture and development], in Edelin (C.) (Ed.), *Naturalia Monspeliensa*, Special Edition, No. A7 (Proceedings of the 2nd International Symposium on the Tree, 10–15 September 1990), Montpellier, 1991, p. 239: "I believe that the notion of production or of recursivity systems in branching topology is more pertinent for present-day botany than solely geometric self-similarity".
8 *Synthèse d'images figuratives d'arbres par des méthodes combinatoires.*
9 Aono (M.), Kunii (T.L.), "Botanical tree image generation", *IEEE CG&A*, 1984, Vol. 4, No. 5, May, pp. 10–34.
10 Private correspondence of 4 May 2001. It refers to Kant's famous statement. Here is the quotation in question: "It is indeed quite certain that we cannot adequately cognise, much less explain, organised beings and their internal possibility, according to mere mechanical principles of nature; and we can say boldly it is alike certain that it is absurd for men to make any such attempt or to hope that another *Newton* will arise in the future, who shall make comprehensible by us the production of a blade of grass according to natural laws which no design has ordered. We must absolutely deny this insight to men". Kant (E.), *Critique of Judgement*, 1790, translated by J.H. Bernard, London: Macmillan and Co. Ltd., 1914, § 75, pp. 312–313.
11 Françon (J.), 1991, art. cit., p. 242.
12 In the early 1970s, the Association for Computing Machinery (ACM) founded an inner special interest group that brought together computer scientists specializing in graphics. This was called the ACM – SIGGRAPH (Special Interest Group in GRAPHics).
13 Here, GERDAT stands for *Gestion de la Recherche Documentaire et Appui Technique* (Management of Documentary Research and Technical Support).
14 A procedural programming language is based on variables, data structures and subroutines, i.e., a series of actions defined by elementary procedures.
15 See Glossary: "object-oriented programming".
16 Smith (A.R.), "Plants, fractals, and formal languages", *Computer Graphics*, 1984, Vol. 18, No. 3, July, p. 2.
17 According to Edelin, some of Hallé's architectural models appeared closer than others and should therefore be regrouped under several "super-models". The so-called "computer" model would make it possible to specify these more appropriately from a botanical point of view. See Jaeger (M.), "Représentation et simulation de croissance des végétaux" [Representation and simulation of plant growth], Computer thesis, Université Louis Pasteur, Strasbourg, 1987, p. 133.
18 Which goes to show that a simulation *by* dynamics is not always a simulation *of* dynamics.

19 Barthélémy (D.), Blaise (F.), Fourcaud (T.), Nicolini (E.), "Modélisation et simulation de l'architecture des arbres: bilan et perspectives" [Modelling and simulation of tree architecture: assessment and outlook], *Revue Forestière Française* [French Forestry Review], Special edition of 1995, Vol. 47, p. 71.

20 SESA: Software and Engineering for Systems and Automata (*Société de services et des systèmes informatiques et automatiques*).

21 "New Images Forum".

22 Reffye (de) (P.), Edelin (C.), Françon (J.), Jaeger (M.), Puech (C.), "Plant models faithful to botanical structure and development", *Computer Graphics*, 1988, Vol. 22, No. 4, pp. 151–158.

23 This concept dates back to a 1935 article by J.H. Priestley, L.I. Scott and E.C. Gillett. A growth unit may include several internodes. Blaise pointed out that "Growth in length of a leafy axis can be broken down into two phases: first, the internodes are created in the meristem, then a certain number of these internodes elongate over a short period of time [. . .] this growth may occur continuously or rhythmically [...] the portion of stem formed during a period of elongation is called a *growth unit*", Blaise (F.), "Simulation du parallélisme dans la croissance des plantes et application" [Simulation of parallelism in plant growth and its application], Computer science thesis, Strasbourg, Louis Pasteur University, 1991, p. 27.

24 Ibid., p. 75.

25 Computer-aided design.

26 These epistemological reflections on the existence of a "module", i.e., on the ability of an elementary sequence of a natural process to mark time for all the other sequences involved, were the subject of an important observation made by Coquillard (P.), Hill (D.R.C.), *Modélisations et simulations d'écosystèmes*, Paris, Masson, 1997, p. 136: "When the relationship between *the time-scale of event occurrences and the time-step selected is highly variable*, it is interesting to select a time management based on events". In other words, simulations managed by clock must be abandoned. This appears to be a conceptual limitation for the production of global models with scale changes. The same authors defined the events approach as follows: "With event-based time management, virtual time progresses from one date of occurrence of events to another. There is no longer a search for events to be processed between time *t* and *t* + (time-step). On the contrary, it is necessary to manage a schedule that stores the events in chronological order. Using this approach, when the model experiences a long period of inactivity, the simulation passes directly to the next significant event", ibid., p. 136. To sum up, we could say that, in an event-based simulation, it is the order that creates the time and not time that creates the order.

27 Blaise (F.), 1991, op. cit., p. 76.

28 See Glossary: "pixel", "voxel".

29 See Blaise (F.), 1991, op. cit., pp. 138 and 146.

30 Barthélémy (D.), Blaise (F.), Fourcaud (T.), Nicolini (E.), 1995, art. cit., pp. 71–93.

31 Blaise (F.), 1991, op. cit., p. 175: "The richness and flexibility of this structuring (which, for that matter go hand in hand with its complexity) give free rein to new applications. After all, let us not forget that it is merely a representation of a botanical reality that is, itself, very complex".

32 This problem of priority management of contemporary events, or those that are considered as such, and therefore of the reliability (or otherwise) with regard to the choice of the sequence of processing these events, is frequent and leads to artefacts in simulations on "multi-agent systems" (see Glossary) in particular. This creates problems of software portability and therefore impacts the verifiability and robustness of these simulations.

33 Elguero pursued his career as a research engineer at the Research Institute for Development (*Institut de Recherche pour le Développement* – IRD, previously ORSTOM), where he worked on epidemiological models.

34 The Poisson process is based on the intuitive notion of destructured random events. This involves making two hypotheses: 1) that events have a constant rate of occurrence (stationarity or homogeneity over time); and 2) that the number of events occurring during two unconnected time periods are independent. A Poisson process can therefore be expressed in accordance with the formalism of renewal processes. A renewal process expresses the law governing the interval of time that separates two consecutive events. In a Poisson process that is expressed in this way, the times between occurrences are distributed exponentially. The theory of point processes (the processes governing the "point events occurring in a haphazard way in space or time", (Cox (D.R.), Lewis (P.A.W.), *The Statistical Analysis of Series of Events*, London, Methuen and Co. Ltd, 1966) is a difficult chapter in statistical analysis and estimation analysis. It developed over the course of the 1960s, in particular in response to the work of D.R. Cox from Birbeck College, University of London. One of its first applications was an analysis of the occurrence of machine breakdowns.

35 His field of specialization was in fact "Physiology, biology of organisms and population biology".

36 See Blaise (F.), Barczi (J.F.), Jaeger (M.), Dinouard (P.), Reffye (de) (P.), "Simulation of the growth of plants", in: T.L. Kunii, A. Luciani (Eds) *Cyberworlds*, Tokyo, Springer, 1998, pp. 81–109, here p. 85: "Daniel Barthélémy defines the concept of automatic flowering in his thesis. He shows that, with all trees, the architecture of a basic model goes through a metamorphosis as it ages which is correlated with the progressive appearance of flowering over all the structure of the plant".

37 Reffye (de) (P.), Dinouard (P.), Barthélémy (D.), "Modélisation et simulation de l'architecture de l'orme du Japon *Zelkova serrata*: la notion d'axe de référence" [Modelling and simulation of the architecture of the Japanese elm, *Zelkova serrata* (Thunb.) Makino (*Ulmaceae*): the concept of reference axis], in Edelin (C.) (Ed.), *Naturalia Monspeliensa*, 2ème Colloque International sur l'Arbre [2nd International Symposium on Trees], 10–15 September 1990, Montpellier, Special edition No. A7, 1991, p. 251.

6 Convergence between universal simulation and forestry (1990–1998)

In this chapter we will see how, starting from 1990, a series of somewhat autocratic or authoritarian decisions led to a further convergence, in particular by leading the CIRAD laboratory (which by that point was well established and flourishing) to join forces with INRA, despite an initial reluctance on the part of the physiological or eco-physiological modellers insofar as simulation was concerned. This epistemological reluctance and the debate that ensued will form the initial focus of this chapter. It is significant that, at the very moment when the various practices had diverged the most completely, it became essential to undertake an epistemological stocktaking and adaptation, which then had to be explicitly set out in the technical publications themselves, far removed from the realm of the epistemologists. The resistance focused principally on the necessity, or otherwise, of using simulation as what might be termed a "maximal model", contrary to the traditional pragmatic-model epistemology that had always promoted the creation of "minimal models". We will see how, based on this episode, the CIRAD laboratory created new combinations of disciplines (after the initial focus on botany, agronomy and computer graphics, forestry was also added) and institutions, enabling simulation to confirm its considerable powers of integration. It was also in the mid-1990s that universal simulation developed its empirical nature to the maximum, even in the eyes of its users, to the extent of creating a direct practice of simulation on simulation (such as, for example, the simulation of bad weather on a simulated forest). I will analyse the precise reasons, and their context, behind this remarkable practice that I propose to call "supra-simulation".

In 1991 Blaise's work thus concluded to a frisson of conceptual excitement in the world of botany, especially among the disciples of Hallé. It took some time, however, for the news to spread in agronomy circles, especially insofar as INRA was concerned. Despite this, an AIP[1] combining AMAP and various INRA laboratories had been launched the previous year. This was a somewhat authoritarian but crucial decision on the part of INRA's management at the time and, in a sense, it would overturn certain firmly rooted habits among the institute's agronomists. This decision by the head of INRA was thus taken largely counter to the general opinion of his staff. Nonetheless, it had the undeniable result of facilitating a rapid and unexpected convergence between architectural simulation and forestry. For many years forestry experts had resorted first to quantitative laws and then to mathematical modelling, largely in the United States to begin with, but later

DOI: 10.4324/9781315159904-7

also in France – especially because of its "breeding ground" of highly qualified X-ENGREF[2] engineers. At the time, for reasons similar to those that de Reffye had set out in the case of coffee plants in 1974, forestry specialists had become more and more interested in approaches dealing with heterogeneous stands through the use of models centred on individual trees. There were other specialized plant-production sectors that used modelling and that could also have adopted AMAP's methods, but this did not occur. The situation was therefore complex. I will not go into detail regarding the involvement and dealings of the INRA managers, including the Director at the time, Jacques Poly, as well as Alain Coléno – who, as we saw, had backed de Reffye during his entrance exam and who had since become Director of the "Plant Production Sector".

In 1991 INRA therefore issued an official call for bids in order to assist in the "development of research on the simulation and modelling of fruit- and forest-tree architectures".[3] This call for bids was made in the context of an AIP that could combine the laboratories of INRA and of other institutions. The researchers who responded to this appeal came, on the one hand, from the INRA Departments of Plant Improvement (Fruit Farming) and of Forestry Research, and on the other from CIRAD's Laboratory of Plant Architecture Modelling. The initiative was extended until November 1993, when a synoptic symposium[4] was held in Montpellier. The collaboration of the CIRAD modelling team was of course openly and keenly desired. Nevertheless, the directors of INRA's Forestry Research hesitated for a long time over whether to collaborate with AMAP. Indeed, for their part, certain dendrometry specialists, along with their arboriculturist colleagues, preferred a different concept of the model to that of AMAP. The debate was all the more bitter since it concerned the very nature itself of scientific activity in the domain of plants: what is plant modelling?

An epistemological dispute between modellers: INRA and CIRAD

In the view of many INRA researchers, it was not essential to use such complex methods as simulation; simple local modelling was enough, without seeking to add complexity from a global point of view when the whole aim was in fact to clarify the phenomena. This was similar to Legay's opinion, and was also shared by most biometricians and ecophysiologists in the 1970s: in their view, the model was a tool for exploration and comprehension; it had a specific purpose by nature and could not be used universally. A model could account for only one or two aspects of reality at the same time, no more. Furthermore, it was based on the interaction between different elements that should, when considered individually, have a well-defined biological meaning. Nonetheless, Jean Bouchon, who was then Director of Forestry Research at INRA, backed the AIP and criticized this epistemological stance as belonging to what he called "the French school of modelling".[5] In so doing, Bouchon contrasted "optimal models" with "maximal models". An optimal model was minimal to the extent that only what was strictly necessary was retained in order to account for a specific plant development. These were the only valid models as far as the French school of modelling was concerned, whereas the supporters of "maximal models" were presumed to uphold the old adage, according to which "he who can do more can also do less". For the adherents of maximal modelling, regional problems could be resolved by means of "successive reductions or degradations" of the maximal model. Simulation, however, was the very essence of a "maximal model". The mis-understanding lay in the issue of knowing what a model should include in order to be valid. What object should be modelled? Is architecture a good subject for modelling? What point of view should the modeller adopt: reductionist or holistic?[6] Should their premise be that the parts explain the whole, or that there is more in the whole than in the sum of all the parts?

This ancient debate, which arises repeatedly in the transitional phases that are typical of the history of formalisms, raises the question of knowing whether or not a certain opacity, or areas of obscurity, are acceptable in models, and whether it is possible to factor unknown relationships, i.e., relationships that the imagination cannot envisage, into the equation. In essence, is it possible to model a relationship that we would not be capable of sketching out with paper and pencil? Should the model always be reducible to a schematic depiction of the processes involved? Or should we resign ourselves to the fact that a model of a living being might be no more than a set of mathematical expressions that might refer, in part and on certain levels, to a fiction and not to an entity with a biological meaning? For the members of the steering committee, the nature of simulation models – which, by then, was very fragmented and potentially fictional at a certain level and for certain components – should not be feared. These different types of models should not be pointlessly set against each other, but instead should be made to meet, made to converge on a common formal ground.

But was it possible to make models that don't have the same epistemic status meet on a common ground? Furthermore, it would not actually be a meeting on a level playing field, since in fact simulation would have a dominant commanding position, right from the start, because of its different nature. This was what some of the INRA dendrometry experts feared when they wondered what explanations simulation could really offer to justify allowing it to occupy such a dominant position in the small community of models. In their view, if one model had an advantageous position compared with others, it must necessarily be because it explained better than those others. Was this the case for simulation?

The second fundamental criticism, which was related to the first, focused therefore on the uniquely descriptive and non-functional nature of AMAP's model. If it was asked, for example, why trees bifurcate, the model would not provide an explanation, even though it was able to simulate a branching tree. Bouchon responded that, once it was known why trees branch, this knowledge of their subtle inner workings could easily be integrated into the structure of the AMAP program. The answer therefore lay in making researchers understand architectural simulation's ability to support any functional sub-model that might arise. Although architectural simulation is an integrator of models, it does not impose an explanation beforehand to unify and simplify the diversity of the phe-nomena: it is not, strictly speaking, a meta-model endowed with a preliminary unifying ontology. Architectural simulation peddles very few physiological hypotheses, since it is based on systematized descriptive foundations on a completely different level from that of the cell or of physiology.

This objection and the response it received demonstrate how reluctant the researchers were to admit the new epistemic function of simulation to the field of modelling. Simulation may well have been a model, but this model was initially conceived without any explanatory aims. Bouchon tried to explain to the researchers that if simulation was a model in the sense that they understood it (a meta-model), then they would be justified in believing themselves to have been wrongly beaten by their CIRAD colleagues. This was not the case, however: it was not their models that were at fault therefore, but instead, more profoundly, their definition for the notion of model. The AMAP "model" was one of the "complicated descriptive models"; they were counter to the "simple functional [or explanatory] models".[7] This was the source of the misunderstanding.

A third criticism can be found in the correspondence between researchers during the three years of the AIP. The INRA modellers questioned what use could be made of a simulation that was unable to "predict the future of a tree that has undergone pruning, or the future of a stand that has been thinned".[8] This problem originated from the stochastic approach: the realism of this approach, in a botanical sense, came at the cost of the merely probabilistic nature of its predictions. The future of an individual could not be predicted with certainty. The INRA dendrometry experts were willing to admit to a certain vagueness in their knowledge of physiological phenomena. But this was necessary in order for a prediction to be possible. If something could not be explained, it should at least be possible to predict it! The AMAP model appeared to permit neither the one nor the other. When expressed in these terms, the criticism was valid. It obliged Jean Bouchon to address the ultimately essentially empirical status of simulation. He replied that, in fact, simulation could be used to "save on field experiments".[9] By this, it should be understood that simulation made it possible to have model organisms for the first time in forestry. These simulated organisms could play the same role as the model organisms replicating the *E. Coli* bacterium in molecular biology, or drosophila in genetics. This was because their rate of growth could be increased tenfold and it was therefore possible to envisage carrying out "experiments". Simulation made it possible to cut loose from the technical constraints inherent in the formation of complex living beings in real life: the long duration, the elapsing of time. For Bouchon, simulation was well and truly a formal model in a new sense, in the sense that it took place among the representative individuals of a living species that had been chosen because they were easier to study than others. Such organisms are usually called "models" because they are assumed to be subject to the same growth phenomena as the others, and because, in addition, they have the quality of making these phenomena more "readable", more accessible to observation and measurement. Computer simulation, in this precise scientific use, replaces an actual "model organism". This was why it is ultimately more than a mere "architectural mock-up"[10] through which one would simply connect explanatory regional models. Simulation becomes a substitute for reality, an object of study in itself, an object of curiosity transposed into the computer, a transfer of as-yet unknown phenomena of living beings into the machine. Precisely for this reason it can give rise to experiments in the same way as an actual plant. It tends to blend in[11] with the area of investigation itself. Simulation is not just a formal shareable ground in the sense that a common language is spoken there. It is a new ground for new experiments. The main aim of a complex simulation that replicates reality is not that of representing the real in an understandable manner (a model for understanding), nor in a directly operational manner (a model for action or decision). On the other hand, simulation requires calibration beforehand in order to replace the real, and to thus become a model for experiment and for use indirectly in an investigation, whether this investigation is aimed at understanding or is intended for use/prediction.

Ultimately, the AIP received rather mixed reviews: the CIRAD researchers (de Reffye, Costes and Barthélémy) had worked intensively over two years to obtain processable data with a view to resolving issues specific to INRA. The INRA researchers, however, had not always played their part. It was at this point that Coléno made a second authoritative and crucial decision: in order to prevent the INRA researchers from continuing to overwhelmingly neglect architectural simulation, and ending up at a later date merely reinventing something that was already available at CIRAD, he considered it necessary to extend the AMAP–INRA link by making it organic. One solution would have been to directly recruit Philippe de Reffye. But Coléno refused to do so: INRA had been incapable of seeing his merits at the beginning of his career; it would have been unfair to steal him from CIRAD so late in the day, at the very moment when he was starting to obtain results. He therefore chose the solution of closer ties between the institutions. In the meantime, at the end of 1993 and also at Coléno's urging, Daniel Barthélémy and Evelyne Costes were recruited as INRA researchers, although both remained employed by AMAP. While the AIP that Coléno had created forged ahead, AMAP continued, almost on its own, its convergence towards the issue of biomass allocation in the plant structure, although the important collaboration in this domain of the dendrometry expert and X-ENGREF Engineer François Houllier – who at the time was a newly recruited teacher and researcher at ENGREF – should already be noted. From 1994, the laboratory began to direct its work specifically towards plant functioning,[12] and the INRA/CIRAD collaboration on AMAP was formalized.

Conceptual and institutional convergence: the CIRAD/INRA partner laboratory (1995)

Following the completion of Blaise's thesis and in the face of the adverse effects of simulation on botanical concepts, a need for clarification and conceptual systematization once again made itself felt. De Reffye noted that each time there was a change of plant, further ad hoc modifications were still necessary in order to make the integrated computer model and its stochastic processes correspond to the new subject under study. In other words, the computer model lacked generality.

Furthermore, AMAP suffered from an absence of internal compatibility between its own programs. In order to resolve these problems, CIRAD recruited two mathematics/computer scientists as research engineers, Yann Guédon and Christophe Godin, who began to develop the AMAPmod software (where "mod" stood for "modelling"), and the AML language (for "AMAP Modelling Language"). AMAPmod software was used to systematically estimate the parameters of the stochastic processes involved in a simulation managed by Jaeger's software. The latter had in the meanwhile been renamed AMAPsim (for "simulation"), and had been rewritten by research engineer Jean-François Barczi, who had integrated concepts of physiological age and of Markov processes.[13] AMAPmod used the AML standardized description language, which in turn was based on the concept of automaton or "axis of reference". In this new formal framework, the simulated plant could be conceived as a "multi-level graph"[14] generated by a stochastic process. This process itself could be formally replaced by a "dynamic probabilistic model"[15] along the lines of a Markov state-transition automaton. As a result of this formal rewriting, the renewal theory became invaluable because it then became possible to resolve the inverse mathematics problem, which involves tracing back to the parameters of the processes once their empirical distributions in the field are known. It became possible to identify (i.e., to quantify the parameters) and validate the model more easily. AMAPmod would later play a decisive role in certain subsequent convergences between the AMAP simulation model and other disciplines, such as mathematical logic, for example. I will return to this issue shortly.

Following another of Coléno's wise but somewhat autocratic decisions, on 1 January 1995 AMAP became an "INRA/CIRAD Partner Laboratory – Plant Modelling Programme". De Reffye remained Unit Director, under CIRAD. He was also made Research Director of the GERDAT department. The Unit comprised 28 permanent staff in all: 14.5 in plant modelling (6 botanists, 1 physiologist, 2 agronomists, 1 agroforestry expert and 4.5 computer and mathematics specialists). Three researchers from INRA formed part of this team. For its part, the computer graphics team included 5.5 members of staff. The computer graphics section was also charged with the new task of adapting the AMAP software to increase its ease of use and thus improve distribution and marketing. As a result, AMAP sold a number of their software products as commercial versions aimed mainly at architects, landscapers and large agricultural businesses. This initial marketing drive, which was also backed up by the INRA and CIRAD management, ultimately required the creation of the post of Administrative Director of the Unit. But, faced with a fall in sales in 1995 compared with the initial fairly flourishing returns of the early 1990s,[16] the auditors of the first external review in 1996[17] ultimately recommended that the unit be split in two. They based this recommendation on the model used at USDA (United States Department of Agriculture), which advised making a clear separation between modelling and the production and marketing of software.[18] In 1995 de Reffye tasked the JMG Graphics company (which changed its name to Bionatics in 2001) with distributing the AMAP software. Neither CIRAD nor INRA ultimately felt qualified to sell software in the long term. In 1996 the "image analysis and remote sensing"

Figure 6.1 Illustration of silver poplar created using AMAP-CIRAD software (1996)
(now *UMR plant botany and bio-computing*). Reproduced by kind permission
of the author.

team, run by Marc Jaeger, comprised 2.5 staff members. Lastly, management and
administration employed a further 5.5 staff. Along with these permanent members
of staff, AMAP had 4 associates, including Claude Puech, who took over from
Jean Françon as Adviser in computer-generated imagery. Thus, as we can see,
AMAP was expanding largely by expedience, by co-working, gradually increasing
its ties and collaborations. This now occurred on a regular basis, in accordance with
external requirements.

The empirical value of simulation

During this period of consolidation (1995–1996), Philippe de Reffye also had a
number of scattered thoughts on the new nature of the models that he was offering
to agronomists. It is interesting to review the spirit of these reflections, since they

seem characteristic of a major epistemological change that was brought about by the emergence of computer simulation, in particular from de Reffye's point of view, as someone who developed and used it in practice. In accordance with the ideas advanced by Bouchon during the AIP, de Reffye was now appealing to what he considered to be the well-established legitimacy of this type of model, contrary to the recommendations of the French school of modelling, which had always favoured and approved special and regional pragmatic models. In a 1996 report compiled by de Reffye in preparation for the external evaluation of the laboratory, after listing the four types of models that could be implemented at that point in the AMAP infrastructure of architectural simulation (statistical model, plant production model, competition model and morphological model), he wrote:

> In the final analysis, all these models complement each other, and clearly a general model that would encompass problems of morphology at the same time as issues of interaction with the surroundings would have an exceptional multifunctionality in its agronomic applications. This is preferred approach in the Plant Modelling Unit.[19]

A little further, we read:

> The choice of a simplified representation of plants is primarily a practical one. Geometric and biometric description of a plant canopy *in situ* is in fact very difficult and is always incomplete. On the contrary, virtual plants are computer objects whose geometry and topology are described completely. It therefore becomes possible to use digital simulation models that make the most complete use possible of the available information [. . .] The choice was made to develop models that were as precise and detailed as possible in order to better analyse the phenomena under examination and, where necessary, to test and reinforce the classic [analytical] models.[20]

Later still, he added:

> It is clear that we must evolve rapidly towards a computer model of the tree that can be used for simulation not just in agronomy, but also in botany, physiology, mechanics and wood quality. The simulation of the tree lies effectively at the crossroads of all these scientific disciplines and enables them, for the first time, to truly communicate between each other.[21]

"Fragmented" modelling (to use my own term) and integrative universal simulation would thus allow a previously unheard-of "multifunctionality" that was no longer to be feared. Such multifunctionality should be contrasted with the single-function nature so often recommended in the epistemologies of models, whether these accompanied a theoretical function of the model (which abstracts and mono-formalizes in order to conceptualize and understand the essence) or, on the contrary, a pragmatic function (which sets a goal, chooses a formal descriptive language and mono-formalizes in order to direct the action). Admittedly, de Reffye rediscovered an interest, which he'd had from the start, for the "laws of plants". But de Reffye sided with the objective measurement of plants in the face of the aprioristic mathematicians and theorists, who were too rushed to truly measure the variability of living phenomena in the field. His biometric side can be seen here. The variability of living beings is present, but it has been both dominated and dealt with by computer simulation, for it has no longer been simply summed up and smoothed over by the global parameters, but instead has been made almost totally reproducible by means of fragmented algorithms. These algorithms could be said to stand in for the laws of plants, as it were, even though the notion of *mathematical model* is lost in this way, and the more modest notion of "digital model" must take precedence. But on this issue, however, de Reffye's terminology remained unresolved in 1996: should one speak of "simulation", of "digital model" or of "computer model"? This indecision proves that the nature of this model was clearly recognized as being different from the others, without its users being able to reuse earlier categories in a stable and satisfactory manner.

Nonetheless, it is clear that there was no longer any need to resort to the old argument of the infinite and therefore almost divine complexity of nature, as was the case in the epistemology of the French school of modelling, and to an even greater extent in the pragmatic epistemologies. On the contrary, and a posteriori, this argument seems to have served to mask an impossibility that de Reffye *decided*, for his part, to consider altogether incidental (see his phrase "choice [. . .] is primarily a practical one"). He no longer felt the need to treat it as sacred, or to interpret it in terms of a taboo or an inherent impossibility due to human finitude: for him, it was simply a case of a momentary technical inability to take sufficient account of information in the models, even though this information was actually available. Since contemporary epistemological iconoclasm was no longer necessary in order to legitimize a posteriori a new technical solution that already existed, this iconoclasm thus revealed, on the contrary, its fragile status as a simple preliminary ontological decision and option rather than as a truly fundamental knowledge that was assumed to systematically impose an uncertain "humility" on the method of models and modelling.

The multifunctionality of simulation, of the "virtual plant", did have an epistemological cost: it brought about a change of epistemic status, which de Reffye also became aware of, at that time. As Bouchon sought to tell the eco-physiologists, simulation is not so much a model as a copy of reality. As such, it plays a role of virtual replication, of an empirical copy of reality, rather than a role of symbolic summing-up or of an abridged set of observational statements (which, for that matter, are accessible on our sensitive scale), or even of a formal model that can be summoned up by the mind for directly cognitive or practical uses:

> The model conceived at CIRAD is original because it is developed on the basis of qualitative knowledge of botany and of plant architecture, to which is added the quantitative knowledge acquired during agronomic experiments carried out at CIRAD's overseas branches or in agronomy research centres in France. Its approach is therefore of an *experimental*, and *not algorithmic*, nature.[22]

In this quote de Reffye was contrasting an algorithmic approach with the experi-mental nature, which is clearly assumed here, of simulation. The algorithmic approach should be understood in a strict sense as the provision of a mathematical model that can be calculated a priori in a series of purely mathematical or logical operations that are homogeneous between themselves and finite in number.[23] This type of modelling is still mathematical. It enables the modelled phenomenon to be under-stood, at least in principle, i.e., it allows the unique mathematical act that is behind the mathematical model to be mentally reconstructed. This type of modelling is the opposite of simulation with fragmented formal support, which no longer promises to increase understanding. Simulation, in turn, produces "digital mock-ups", "computer objects"[24] that are compact and opaque,[25] on which, however, virtual experiments can be carried out. Simulation thus becomes a distinct ground for experimentation. It must be clearly understood that this experimentation is carried out by proxy, as though in the field, once the simulation model has been correctly calibrated in the actual field, from a quantitative point of view. In this way, observations are made only on individuals, on copies of single objects, which themselves are also only one, whose characteriza-tion and generality remain problematic. This singleness likens the simulations to actual experiments. Their reproducibility (as long as they use the same parameters and the same "seeds" for the generation of pseudo-random numbers), however, likens them to conceptual arguments or to thought experiments.[26] In this way, a number of seemingly contradictory epistemological theses were upheld in this regard.[27] At that time, de Reffye considered for his part that it was entirely justifiable at that point to speak of "virtual agronomic experiments".[28]

Supra-simulations

The best illustrations of AMAP's tendency to use architectural representations in all their complexity as mock-ups or models to be used in virtual experiments are to be found in works on the simulation of "radiative climates". Such work originated in the context of a specific and concrete agronomic requirement. At the end of the 1980s Jean Dauzat, an Agricultural Engineer with a Master's degree in Ecology, joined the team. His task was to develop the agronomic applications of the simula-tions. In 1990, during an internal ATP (*Action Thématique Programmée*[29]) within CIRAD, Dauzat suggested using architectural simulations in order to estimate radiative transfer in an oil-palm plantation. His idea was to precisely evaluate the intercepted light, the light transmitted below the canopy and also the directional reflectance. The agronomic aim was to optimize the density of associated crops (i.e., the underplanted crops or intercropping). The classical mathematical mod-els used for evaluating radiative exchanges under a plant canopy, however, used extremely simplified plant representations. As with traditional biometric repre-sentations, the leaf distribution was considered to be random in such models, and it was the average effect of the plant canopy on sun exposure over the course of a day that was sought. But these models did not take account of the observed and measured heterogeneity of illumination reaching the ground. It was therefore not possible to ensure optimum planting density.

Between 1989 and 1994 Dauzat, assisted by Lecoustre, used precise architectural measurements taken in Côte d'Ivoire in order to construct realistic mock-ups of palm trees. First, he confirmed the heterogeneity of sun exposure by using captors to carry out measurements of the radiation transmitted to soil level, as well as measuring the diffuse radiance under the palm trees. Then, by dividing the surface area into elementary triangles, he found that the average rate of transmission per triangle obtained by simulation using the AMAPpara software was very close (to within ±1 of 10%) to the measured rate.[30] The simulation therefore closely *replicated* the heterogeneity of the radiance. Lastly, using the same mock-ups, Dauzat simulated the total radiative climate, with its re-diffusions and re-interceptions, by using the computer to follow the history, or in other words the evolution, of a large number of incident beams originating from the sun throughout the day. In order to obtain a stabilized climate, the simulation required a large number of these virtual "ray tracings": between 1 and 2 million in the case of a small palm plantation scene. Dauzat observed: "This method is very time-consuming as far as calculation times are concerned. On the other hand, this mechanistic approach is very rigorous and allows the radiative exchanges to be described in detail".[31] Dauzat had thus demonstrated that the mapping of the total radiation balance of a plantation could be obtained in detail if its architectural details were reproduced by computer beforehand. This was a similar method, which made it possible to provide an image of the reflectance, as seen from above, of a complex plant scenario; this explains the connection that was soon made with the problems of interpretation of images produced by remote sensing (aircraft, satellites, etc.). In this context, from a conceptual point of view, the computer simulation of architectural botany thus made it possible to bypass the problems of non-calculability by reconnecting with the roots of digital simulation:

> The ray tracings model that we have developed is aimed at the most precise simulation possible of canopy reflectance. It is too complex to be inversed. By means of simulation experiments, however, it is possible to establish a relationship between a measured signal and the various characteristics of the canopy. The relationships thus obtained can then be inversed.[32]

With this use of the computer, the mock-ups – since they are themselves the result of earlier simulations – become in turn the ground for another type of simulation: illuminance simulation. At the end of their construction process, their aspect is therefore not yet globally summed up by a model or formula. The model is then used once more, as is, with all its roughness and with its heterogeneity preserved.

It is at this point that architectural simulation becomes an empirical ground in its own right, by giving rise to what I will call in this case a *supra-simulation*. Dauzat pointed out many times, for that matter, that this simulation by "computer mock-up" quantitatively contradicts the usual condensing models. Such simulation by mock-up could therefore be used to *test* the mathematical models. With reflectance simulation, the mock-up could also be used to *interpret*, albeit inductively, remote sensing images. In other words, the mock-up could be used to determine what type of plantation and what species

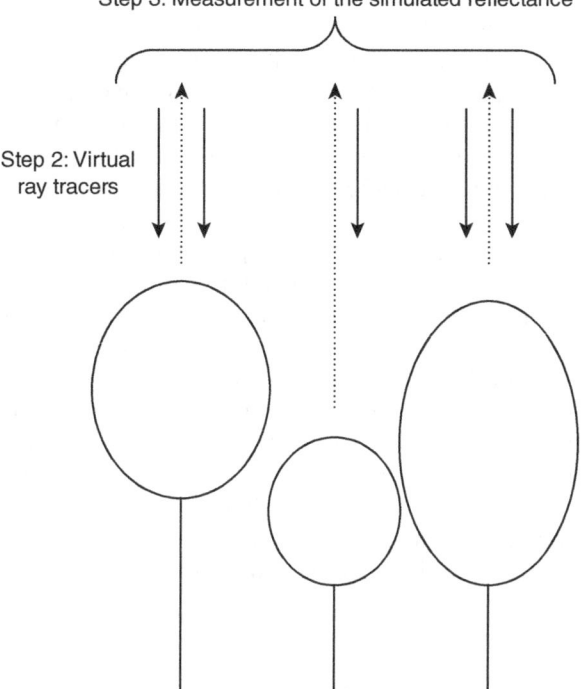

Step 3: Measurement of the simulated reflectance

Step 2: Virtual
ray tracers

Step 1: Heterogeneous forest simulated using
AMAP software

Figure 6.2 The three steps of supra-simulation: the case of reflectance simulation.
Author's own representation.

of trees are represented on an image. During the 1990s Dauzat persisted with
the conception and refinement of radiative exchange models by means of 3D
architectural simulations. Until at least 2002, several theses were presented
on this subject and using this approach.[33] In 1995 Dauzat began to partici-
pate in the PNTS (*Programme national de télédétection spatiale*[34]), where,
together with INRA's Bordeaux and Grignon Bioclimatology Units, he proposed
a "study on the reflectance of Maritime Pine stands in the Landes forest".[35] He
also contributed to the interpretation of radar data that had been acquired earlier
by CIRAD-Forest during an airborne measurement campaign carried out on a
eucalyptus plantation in the Congo.

Nonetheless, at the end of the 1990s, an important question arose for
AMAP: could this same use of simulation, this sort of *supra-simulation* or
simulation on simulation, be developed in the domain more directly concerned
with evaluating the mass of matter produced (wood, fruits, etc.), which is a
major focus in agronomy? What came next showed that, although this idea of

extending a simulation that had been first conceived as an empirical ground for other problems had dominated the views of de Reffye and his team in the early 1990s, it turned out to have been somewhat over-optimistic even though it was not wrong in principle. Two hurdles very rapidly emerged over the subsequent years. First, a technical hurdle; the production of a purely architectural (wire) "virtual plant" already required such enormous memory resources that the speed of calculation was significantly affected. Further overloading this mock-up with extremely data-heavy physiological modules, unlike Dauzat's relatively lightweight "ray tracer"[36] modules, led to practical problems of calculability from the start, even for the most powerful computers – as had already been seen in 1993 with Houllier's preparatory work for the AIP.

Then there was a methodological hurdle. In 1996 de Reffye aimed to actually carry out *virtual agronomic experiments*. The history of agronomy had demonstrated for its part, however, that there could be no useable experimentation without clear distinction between the controlled and uncontrolled factors. It is necessary to carry out controlled experiments. As R.A. Fisher's founding approach had demonstrated, there must be a certain interaction between the experimenter and the field entries; their random distribution should at least be controlled and known, even if their value cannot be controlled.[37] The constructed opaqueness of the simulation, which was a sign of its realism and complexity, quickly

Figure 6.3 Five simulations of Araucaria at different ages created using AMAP-CIRAD software (2003), UMR plant botany and bio-computing. Reproduced by kind permission of the author.

became an obstacle in this case: the simulated links had to a certain extent to be disentangled, step by step, in order to see how the optimizations operated. It was necessary to return to a certain sort of comprehension, of condensation, at least insofar as the computer was concerned. It could be said that, even if it was not always necessary for the user, it was necessary on the contrary *that the computer should understand, at least a bit, what it was doing* if the simulations were intended to be used in particularly complex optimization problems, i.e., by applying physiological effects to growth. This was the main reason behind the third convergence that AMAP progressively and fairly deliberately underwent from 1998, after its convergence with computer graphics and its convergence with forestry: this was its convergence with the mathematical graph theory and with automation and, as a result, the belated reconnection with Lindenmayer-type logical simulation. After its emergence in fragmented simulation, where calculable mathematics no longer took priority, it became clear that the representation of plants must be "remathematized", so to speak, even though, contrary to what might at first be easily believed, this was not for essentially theoretical reasons. This was the main driver behind the third convergence that AMAP undertook starting in the late 1990s.

Notes

1 An AIP (from the French *Action Incitative Programmée*), is a type of INRA management policy aimed at setting up collaborations (both internal and with external laboratories) and stimulating funding for projects.
2 Engineers from the *École Polytechnique* who have completed their postgraduate practical training or internship (*école d'application*) at the French National School of Forestry (ENGREF: *École Nationale du Génie Rural, des Eaux et des Forêts*).
3 See Bouchon (J.), "Préface", in J. Bouchon (Ed.), *Architecture des arbres fruitiers et forestiers* [Architecture of fruit and forestry trees], Proceedings of the symposium of 23–25 November 1993 (Montpellier), Paris, INRA-éditions, "Les colloques" ["The symposia"] collection, No. 74, 1995, p. 7.
4 The symposium of 23–25 November 1993 was the subject of two publications: Bouchon (J.), 1995, op. cit. and Bouchon (J.), Reffye (de) (P.), Barthélémy (D.) (Eds), *Modélisation et simulation de l'architecture des végétaux* [Modelling and simulation of plant architecture], Paris, INRA-éditions, "Science Update" collection, 1997.
5 Bouchon (J.), 1995, op. cit. p. 11.
6 In the "holistic" point of view, as Bouchon notes (ibid.), it is assumed that the interaction between the parts creates effects that cannot be foreseen on the basis of the individual parts themselves, nor of the simple sum of the parts' behaviours.
7 Ibid.
8 Ibid., p. 14.
9 Ibid.
10 Ibid., p. 23.
11 Even in 1997, this confusion was still unacceptable for Jean-Marie Legay. Cf. Legay (J.M.), *L'expérience et le modèle. Un discours sur la méthode* [Experiment and model. A discussion of methods], Paris, INRA Éditions, 1997, p. 55: "in any case, the model does not make it possible to dispense with experiment. On the contrary, it suggests and organizes experiments; it may save carrying out a few, but it also proposes others. Of course, even *simulation does not replace the experiment; it is not an experiment*, it makes it possible to detect absurdities, to designate feasibilities. Just because a calculation is possible does not mean it is pertinent. Only experimentation will give a meaningful verdict". Emphasis

added. For Legay, simulation maintains an essentially heuristic role, in particular because it retains the nature of a calculation and can only be a means of exploration of a model.

12 Private correspondence with François Houllier dated 31 October 2000.

13 From the name of the Russian mathematician Andrei Markov (1856–1922). A Markov process is a memoryless stochastic process (at each instant, a random event occurs based on a given state): the probability of an event at time $n + 1$ depends solely on the state of system at time n. This type of process can manage a hierarchy of states and can therefore be produced in the form of a state-transition automaton whose transitions are weighted by the probabilities of occurrence of events.

14 Bouchon (J.), Reffye (de) (P.), Barthélémy (D.), 1997, op. cit., p. 146.

15 Ibid., p. 187–191.

16 Sales had fallen to 600 000 French Francs per year, compared with the annual 1.2 to 1.5 million Francs that the Unit had earned from software sales until then (with a peak of 2.5 million Francs). During the same period, for its overall working costs (staff and equipment), the Unit required 12 million Francs per year. The Unit was therefore far from being able to meet its costs through the sale of software.

17 This system of external review or collective evaluation is carried out every five years and stimulates research and promotion, as well as producing recommendations. It is based on an analysis carried out over six months by external contributors, and gives rise to debates and publications. The system was set up by Hervé Bichat based on the evaluation model used in the major Agronomic Research Institutes, such as IRRI (International Rice Research Institute, an American institute set up in 1971, aimed at promoting research for development in agronomy; 15 other IRIs were later established following this model). This information is based on private correspondence with Henry-Hervé Bichat, dated 20 June 2001.

18 Pavé (A.) *et al.*, Première revue externe de l'unité de modélisation des plantes [First external review of the Plant Modelling Unit], Montpellier, CIRAD, 1996, p. 33.

19 Reffye (de) (P.) *et al.*, (Eds), Document préparatoire à la revue externe de l'unité de modélisation des plantes, CIRAD, Montpellier,1996, p. 67.

20 Ibid., p. 73.

21 Ibid., p. 90.

22 Ibid., p. 115. Emphasis added.

23 See Glossary: "algorithmic simulation".

24 Ibid., p. 115.

25 See Di Paolo (E.A.) *et al.* "Simulation models as opaque thought experiments", in Mark A. Bedau *et al.* (Eds), *Artificial Life VII*, Proceedings of the 7th International Conference on Artificial Life, Cambridge, MIT Press, 2000, pp. 497–506.

26 I will return to these issues in detail in the conclusion.

27 For an initial review, see Varenne (F.), "What does a computer simulation prove?" in N. Giambiasi, C. Frydman (Eds), *Simulation in Industry*, Proceedings of the 13th European Simulation Symposium, Marseille, 18–20 October, SCS Europe Bvba, Ghent, 2001, pp. 549–554.

28 Reffye (de) (P.) *et al.*, 1996, op. cit., p. 12.

29 Scheduled research initiative.

30 Dauzat (J.), "Radiative transfer simulation on computer models of *Elaeis guineensis*", *Oléagineux* Vol. 49, No. 3, 1994, p. 86.

31 Ibid., p. 88.

32 Reffye (de) (P.) *et al.*, 1996, op. cit., p. 134.

33 See the theses by Pierre Guillevic (1999) and Delphine Luquet (2002), for example. Dauzat was a member of the thesis panel and his work was used as a reference. Indeed, his work served as a general framework for reflection and as an assumed starting point in Delphine Luquet's thesis on "The hydric tracking of plants by thermal infrared". Luquet was engaged in 2003 as a researcher with CIRAD-AMIS (*Amélioration des Méthodes pour l'Innovation Scientifique* – Improvement of Methods for Scientific Innovation) in 2003.

34 The National Programme for Space-based Remote Sensing.

35 Reffye (de) (P.) *et al.*, 1996, op. cit., p. 135. In fact, in this specific practical context, Dauzat was already obliged to abandon his needle-by-needle simulation of pine trees: the ray-tracing simulation would have required a prohibitive amount of time, contrary to what had been needed for the small number of palms. He was therefore obliged to remodel "leafy growth units" in a conical and simplified manner, thus accounting for the large numbers of pine needles. *This empirical method was therefore not valid for all trees in practice, because of the power and memory limitations of the mid-1990s computers.*

36 This model later also revealed limitations, in particular insofar as calculation speed was concerned. From 2000, in the context of the INRIA convergence – which will be discussed later – two INRIA researchers, Cyril Soler and François Sillion, assisted by Frédéric Blaise (AMAP), adapted the algorithms obtained through computer graphics in order to more efficiently simulate the radiative light energy under a plant canopy. This light radiosity was in fact a longstanding issue in computer graphics, when the scenes must be made realistic from the point of view of their lighting through the use of fairly well-developed rendering techniques. In the research report by Soler (C.), Sillion (F.), Blaise (F.), Reffye (de) (P.) (*A Physiological Plant Growth Simulation Engine based on Accurate Radiant Energy Transfer*, Le Chesnay, INRIA report No. 4116, February 2001), the authors chose a hierarchical technique whereby equivalent reflectances and transmittances are calculated for the main tree structures on a global scale. The report showed that refinement down to the level of the individual leaf (by means of precise calculation) was possible through hierarchical instantiation, i.e., by descending lower down the scale of data classes. But this required even more substantial calculations. In 2002, parallel computers began to be used for this purpose. The important point is that, as with the case of discretization into voxels in Blaise (1991), a purely mathematical and algorithmic computer-graphics technique could be used to indirectly resolve problems specific to natural science. In this case, it is in the context of calculating an energy value on a simulated mock-up. *The science of computer graphics was therefore not necessarily lagging behind experimental, physical or biological science; indeed, it could sometimes be used as a formal tool for objective simulation or for calculation.*

37 See Glossary: "experimental design".

7 The remathematization of simulations (from 1998 onwards)

The previous chapter suggested that an all-simulated solution was no longer tenable in cases where the aim was to integrate the plant's physiology into the computer simulations. The CIRAD laboratory, which in the period from 1998 to 2002 had been headed by François Houllier, had produced work tending to simplify the original simulations without relinquishing their underlying principles, but at the same time introducing new mathematical sub-models that – at certain stages and in a simplified manner – would deal with the physiological functioning of the plant. Starting from this initial simplification, de Reffye persevered, aided first by the expertise in automation of a laboratory in China, and later by French researchers from INRIA. The establishment of the joint Greenlab CIRAD/INRIA/École Centrale Paris[1] team ultimately formalized the entry into this "remathematization" phase, as I call it, resulting essentially in a belated convergence between CIRAD's approach and the school of algorithmic simulation inspired by Lindenmayer. We will see, thus, that a relatively mature simulation calibrated on a large number of data and, in particular, the implementation of this simulation by computer, provided a formal middle ground on which it was possible to organize in return a simplification and a symbolic standardization – or, in other words, a remathematization. It is significant that, in this context, it was not a concern for conceptualization or comprehension imposed on simulation as a result of outside demand from theoretical biologists, for instance; instead, essentially for reasons of feasibility (regarding calculation times and memory capacities) and of verifying the simulations, it was the computer itself that ultimately needed to understand what it was doing, so to speak.

On 1 January 1998 François Houllier took over from de Reffye as head of AMAP. From INRA's point of view, his position was Director of the "CIRAD-INRA Partner Laboratory for Modelling the Architecture of Forest Trees". On the CIRAD side, he was named head of the "Plant Modelling Programme". This decision was also to a large extent backed by Coléno. Nearly half of de Reffye's time was consumed with various administrative duties (such as the management of GERDAT), and he had been running the laboratory for 15 years by that point: he was rather weary of it all. Above all, he missed being able to apply himself full time to research, especially when he could see that promising alliances were on the point of producing results on a conceptual level: he couldn't see how these

DOI: 10.4324/9781315159904-8

could actually come into being without his constant support. He therefore wished to be free of these constraints in order to dedicate himself once again to research. For de Reffye, Houllier's nomination was therefore a golden opportunity.

The first mixed structure-function model: "water efficiency" (1997–1999)

De Reffye's relief was all the greater because, in the meantime, in 1996, a momentous conceptual innovation took place in the modelling offered by AMAP. Although the first attempt at pairing architectural simulation with ecophysiology had yielded mixed results during the AIP, and it looked like the limits of calculation would rapidly be reached, de Reffye came up with the idea of using a simple and classic law in ecophysiology: the law of water efficiency. This phenomenological law, which had been known to ecophysiologists since the 1930s, postulates that the matter created during photosynthesis is continuously proportional to the plant's transpiration. This is true in cases where the plant is not subjected to hydric stress. At first glance, however, the transpiration itself depends entirely on the architecture. Since the AMAPpara software accounted for the primary growth schedule, including its actual sequence and its topology, if it was possible to evaluate the transpiration at each stage, then it should also be possible to precisely evaluate the production of matter itself. All that would remain to be done would be to allocate it differentially in order to simulate the secondary growth of the organs. This was all the more feasible since local allometry laws were available for each organ. For de Reffye, this provided an opportunity to draw on ecophysiological knowledge, but without entering into the numerous conflicts regarding regional models that enlivened the world of ecophysiology. At the time, a number of French ecophysiologists were up in arms against this new intrusion, which they felt was untimely and lacking respect for the complexity of the phenomena under consideration. In general, when it came to modelling, other than the classically biochemical reductionist quantification approaches on a more integrated scale, the ecophysiologists generally knew only of modelling based on processes that used functional compartments models. Some ecophysiologists, however, such as Pierre Cruiziat, supported de Reffye's approach. Cruiziat recognized the need to improve on the flux-centred physiological models in order to "spatialize" functional tree models.[2]

From 1997, together with Frédéric Blaise, François Houllier, Thierry Fourcaud and Daniel Barthélémy, de Reffye therefore adapted AMAPpara so as to integrate this water-efficiency law.[3] The new model was effected in cycles that can be broken down into three steps:

1   The architectural growth driver determines the primary growth of the axes and organs; the matter created in the preceding cycle is allocated to the leaves and internodes; in order to do so, each organ is treated as a sink for matter; the laws of allometry are used to make the organs grow. All the matter produced is consumed.

2   The calculated geometric structure makes it possible to quantify transpiration by taking account of the architecture of the hydraulic network as modified by the preceding growth. The notion of hydraulic resistivity (using the classic analogy to electricity) then allows the resistances to be summed. This makes it possible to analytically calculate the distribution of the hydric potential.

3   Lastly, the growth of the annual rings of the axes and of the fruit is calculated on the basis of their sink strength and their expansion. The increase in thickness of the axes is brought about by a uniform deposit of matter along the length of the branches.[4]

On the basis of this work, de Reffye and Houllier began to relinquish the simulation-only approach. Instead, they paired a stochastic simulation approach (which still embodied a macroscopic manifestation of the plant's genetic program) with a functional approach in which the mathematical relationships could be expressed analytically at each step. They demonstrated that it was possible to write a "recursive formula linking a tree's growth (transpiration, assimilation and allocation) to its internal structure and its aerial and root architecture":

$$Q_n = \frac{k.\Delta\Psi_n}{\sum_{j=1.2}\left(r_j A_j + r_j{}' B_j \left(\frac{Q_{n-1}}{f_{jn}}\right)^{1+\alpha} + \rho_j C_j \left(\sum_{i=1}^{n-1} \left(\frac{Q_{i-1}}{f_{ji}}\right)^{\frac{1+\alpha}{2}} \right)^2 \right)} \cdot Q_{n-1}$$

where:

$Q_n =$	quantity of dry matter produced at step n;
$\Delta\Psi_n =$	difference in hydric potential (between leaves and soil) at step n;
$k =$	water efficiency;
$j = 1$	for aerial parts;
$j = 2$	for underground parts;
$f_{jn} =$	number of active growth units for water assimilation ($j = 1$) and supply ($j = 2$);
$r_j =$	resistance to transpiration (leaves) or to water absorption (root hairs);
$r_j{}' =$	resistance to circulation of xylem sap in the terminal growth units;
$\rho_j =$	resistivity of sap-conducting growth rings;
$\alpha =$	growth-unit morphology parameter;
$A_j, B_j, C_j =$	parameters depending on the plant morphology and the rules of allocation of the assimilates.[5]

The simulation was still governed by the fragmented model of architectural growth at each stage. But at each of these stages, certain decisive variables were no longer subject to probabilistic construction, as was still the case in the AMAPsim software.

The variable Q_n, for example, was calculated analytically in accordance with the results of the architectural simulation at each stage. The plasticity of the architectural models (which for that matter retained their determinism within a stochastic framework) was thus controlled by the architecture itself, in its relationship to the water efficiency. Thanks to the phenomenological law of water efficiency, which is a validated mathematical model, we see the beginnings of a feedback effect from the architecture on the production by photosynthesis, and of a feedback from this production on the expression of the architectural model (its speed). In the end, we have the beginnings of a regulating model where the architectural model is controlled by a process-based production model and, in return, controls it. Thus we can see that it was the convergence towards agronomy and its problems regarding controlled production that brought about a return to comprehensive mathematical modelling. This approach, which in a way remathematized something that had previously been fragmented, also allowed improved verification of the modelling program. This issue had been a constant matter of concern for de Reffye. He was already aware of it in his 1979 thesis: it should be recalled that he had set himself the task of finding a tree that could be entirely analytically calculated in order to verify that the stochastic simulation of this simple tree would produce the same results.

Ultimately, because it seemed much less heavy (regarding the number of elementary computation steps) and easier to calibrate than the mixed model proposed during the AIP, this model, which had been grafted onto AMAPpara and was initially named AMAPhydro (in 1999) and subsequently AMAPagro, underwent extensive validation testing on the case of the cotton plant to begin with. The qualitative and quantitative results were satisfactory, and the model was considered to have been validated. It was the subject of a number of important articles, in the *Agronomie*[6] journal in particular. In this context, by "validation", de Reffye's team meant "the complete reconstruction of the plant, cycle by cycle, by a biomass produced and distributed in a three-dimensional architecture using parameters measured or calculated from experimental data".[7] Nevertheless, along with this return to mathematical models that was mainly brought about by the need to calculate mixed models, we know that AMAP was seeking to reinforce and refine its formalisms by giving them a less improvised appearance, in particular through the work of Godin and Guédon. From the early 1990s, this formalization work contained the beginnings of a possibility of more direct comparison, or even of closer ties, with the logical simulation techniques of the Lindenmayer school. This merger did in fact take place, however AMAP was not alone in taking steps towards making this reconciliation possible. It took the gradual emergence, albeit somewhat belatedly, of a concern for high-quality graphics and for attention to botanical, geometric and stochastic detail in the formalist Lindenmayer school for the reconciliation to occur. This shift in focus itself has its own history. In order to understand this decisive convergence between the school of L-systems and the school of fragmented modelling and architectural simulation, I must briefly recount the earlier stages of the evolution of L-systems that had taken place since the period where we last left it.

The parallel evolution of algorithmic simulation: 1984–1994

During the 1980s, L-systems remained a subject of great concern for computer scientists, mathematicians and linguists following the appropriation I mentioned earlier. Nonetheless, from 1976 to 1984, Paulien Hogeweg, and later Alvy Ray Smith, both independently tackled the difficulty of immersing a formal grammar in a geometric space. The obvious goal was to make the logical model likely to be suitable in the long run for quantitative validations of field measurements. L-systems were still largely used in this area for theoretical concerns. However fractals – essentially the mathematics of the concrete – were having a significant impact in the rapidly expanding computer graphics circles of that time, even though in the case of plant morphogenesis their direct use was limited to a few species of fern. But although Smith made fractals more flexible in the graftal solution he proposed, they seemed to remain overly rigid. For that matter, his solution was never really espoused. Another approach was more fruitful. Starting in 1971, Andrew L. Szilard, a computer professor from the University of Western Ontario, became particularly interested in L-systems. In 1979, just when fractals were starting to become popular, he and his student R.E. Quinton published an article in an undergraduate journal in which he demonstrated that L-systems could generate fractal curves.[8] The logical formalism of L-systems was thus shown to be at least as flexible as the mathematical approach using fractals. The idea was to include rules of rotation on a plane, as well as classic rules of symbol rewriting, in the L-system axioms. It became clear that L-systems no longer needed additional geometric interpretation: the L-systems themselves generated a geometric diagram on a plane, although this diagram remained discretized and relatively inflexible from a formal point of view.

In 1985 this little-circulated article caught the attention of Przemysław Prusinkiewicz, a professor in computers and computer graphics at the University of Regina in Canada. Prusinkiewicz had followed an unusual career path. He was Polish by birth and had studied computer science at the Warsaw University of Technology between 1970 and 1979, before becoming professor at the Algiers University of Science and Technology from 1979 to 1982. His initial specialism was in problems of error tolerance in numerical calculations. While in Algiers, he was given an opportunity to join the University of Regina. Once he had access to adequate material (Silicon Graphics workstations), Prusinkiewicz was able to direct his attention to computer graphics. It was in this way that he eventually heard of the work carried out earlier by the students and colleagues of Seymour Papert at the MIT Artificial Intelligence Laboratory. Let us briefly summarize the essence of this work: the laboratory was founded by Papert and Marvin Minsky in the mid-1960s. Papert was a mathematician and had worked with Jean Piaget in Geneva. His mathematical approach was therefore tinged with constructivism. One of his first projects was the development of a type of programming language that, using simple commands for movement addressed to "turtles" (which were initially mobile robots on a flat surface), simulated or assisted in a step-by-step constructive cognitive learning process. To do this, he returned to and tested (on physical models to begin with) Piaget's core intuition that concepts have a sensorimotor source. The programming language was named LOGO in 1967. It was based on the LISP language, which had been created in 1960, and was initially used to simulate animal behaviour. But, in 1975, Harold Abelson and Andrea di Sessa, who at the time were both researchers in the same laboratory as Papert, used LOGO to develop what they called a "turtle geometry". Abelson and di Sessa also openly identified with constructivist intuitionism, especially in mathematics.[9] For them, it was a question of combining constructivist artificial intelligence and computer-graphics techniques. The value of the LOGO graphical commands resided in the fact that they did not rely on a system of global coordinates. The geometric properties were "intrinsic to the figures rather than imposed by a reference infrastructure".[10] The construction of shapes was therefore very easy to program.

In 1982 Abelson and di Sessa published their synoptic work, *Turtle Geometry*, with MIT Press. Prusinkiewicz first heard of it in 1985, at the same time as he learned of Szilard's earlier work. He took inspiration from both sources and came up with the idea of combining the flexible LOGO-type approach with the formalisms of L-systems in order to create a system for the graphical representation of complex botanical shapes. In 1986 he published a first article on "Graphical applications of L-systems". He aimed to systematically interpret L-system rewriting rules as a series of commands controlling the movements of a turtle, the behaviour of which he considered to be entirely analogous to the apical meristem[11] of a plant. He then contacted Lindenmayer, with whom he collaborated for some time. Indeed, Prusinkiewicz attempted to convince Lindenmayer that the use of L-systems for computer graphics made it possible to deal more simply with his own initial concerns in biological morphology, the very concerns that had first led him to L-systems in 1968. In particular, the use of the turtle made it possible to avoid the laborious use of linear bracketing. The Silicon Graphics IRIS 3130 workstation that Prusinkiewicz used at Regina had the advantage of truly spatializing L-systems and making them more accessible to intuition. Rather like Ulam in his day in his dealings with von Neumann, Prusinkiewicz promoted a spatialization of the formalism that Lindenmayer had initially hoped could be avoided or bypassed with the aid of theorems and structural linguistics. In so doing, Prusinkiewicz helped L-systems enter a phase during which they stopped being purely theoretical models.

In 1987 Prusinkiewicz and Lindenmayer gave a joint statement at the first conference on Artificial Life, organized by Christopher Langton at Santa Fe.[12] Prusinkiewicz demonstrated that the constructivist approach using the formalism of the turtle possessed another remarkable advantage: it made it possible to introduce L-systems to three-dimensionality. Indeed, it can be considered that the turtle uses vector products to undergo vector rotations in space and thus change its direction. During this conference, Prusinkiewicz and Lindenmayer presented several fairly realistic images of plants. This collaboration soon came to an end, however, with the death of Lindenmayer in 1989.

During the years that followed, Prusinkiewicz and his University of Regina computer graphics students developed what he called a "Virtual Laboratory" in botany. In reality, however, he retained his computer-scientist approach: the virtual laboratory, which he had hoped would be a decisive aid for botany, proved to be essentially a large database of visual data, hierarchized by categories of objects. It was conceived as an attempt at object-oriented programming based on his first software in C language in 1986: *Pfg* (short for *Plant and fractal generator*). The main function of the program was to present scientific information visually. Prusinkiewicz asserted that "new experiments"[13] could be carried out using this program, but he was thinking in terms of pedagogical experiments: Prusinkiewicz considered that – in the line of approach adopted by Piaget, Papert, Abelson and di Sessa – there was a continuity between scientific experimentation that produced truly new information and the type of learning experiments typical of students. In 1991 he was named professor of computer science at the University of Calgary, and it was there that he pursued the development of his software platform.

Prusinkiewicz's initial work came to the attention of the CIRAD team, in particular through Françon and his doctoral students, Jaeger and Blaise. In addition, the richly illustrated book *The Algorithmic Beauty of Plants*, written in collaboration with Lindenmayer just before his death, soon increased Prusinkiewicz's renown. Both Françon and de Reffye, however, considered that the book did not pay real attention to in-depth botanical knowledge. Once again, it was a top-down modelling and was thus rather theoretical, highlighting the formal merely for the sake of it. The virtual laboratory was not based on sufficiently numerous and generalized botanical concepts to be able to match its ambitions, i.e., to have a generic enough nature to be able to replace the reality found in the field in certain innovative experiments. The fact that Prusinkiewicz's software was mainly disseminated in computer graphics circles merely confirmed this view. For his part, Prusinkiewicz was aware of de Reffye's work, but he initially considered it somewhat makeshift from a formal point of view, although he later changed his mind after the remarkable convergence that would take place between AMAP and his own approach.[14] Nonetheless, this convergence was not exactly Prusinkiewicz's doing, even though several meetings took place from 1995. His fundamental epistemology prevented him from doing so. His aim was primarily to apply computer and algorithmic concepts in an attempt to directly improve the understanding of morphogenesis phenomena in living beings.

Thus, in a spirit that was ultimately in keeping with Rashevsky and early theoretical mathematical biology, Prusinkiewicz realized that it was necessary to find a single formalism to achieve comprehension. For Prusinkiewicz, as for a number of his students and followers, computing merely increased the formal arsenal of mathematics. There was an unbroken continuity between the two in the epistemic function that was traditionally attributed to formalisms. In his point of view, the mathematicism of yesteryear should be retained, but must be transformed to a certain extent into "computationalism": it was necessary to dump the categories of computing from top to bottom, and then see whether anything of the real world could be explained in that way. According to Prusinkiewicz, it was necessary to go from formalisms to the laws of nature, whereas for de Reffye the opposite held true: first carry out the measurements, and then improvise the first formulations of the laws, as necessary. Formal refining could come later. Formalisms that were tackled beforehand could always be twisted one way or another in order to give the impression of more or less adhering to the laws.[15] For de Reffye, this did not mean that it was not necessary to seek the best formalism. But this could only be done on the basis of an empirical model incorporating measured laws. Simulations must act as an empirical testing ground for the mathematical models, which remained more desirable than ever.

During the 1990s, Prusinkiewicz's laboratory learned some lessons from AMAP and from what Prusinkiewicz called *empirical models*, as opposed to *causal models*:[16] he made the logical system much more flexible and implemented probabilistic laws in the rules of formal ramification. The fact remains, however, that the need or rather the obligation of actually calibrating these more complex logical models did not come immediately from a computer laboratory, such as Prusinkiewicz's, but instead originated quite understandably from a forest-sciences faculty – that of the University of Göttingen, to be precise.

In 1994 Winfried Kurth, a mathematician and computer scientist, was working on a program to model "the dynamics of change in forest ecosystems".[17] For his doctorate, Kurth had first worked on modelling natural shapes using formal L-system type grammars. He was therefore perfectly aware of Prusinkiewicz's most recent work. When he arrived at the Göttingen faculty, however, the stakes were different from those normally encountered by computer scientists and computer graphics specialists: it was necessary to make these models capable of being calibrated and used for forest ecology analyses. According to the Göttingen researchers at the time, however, if the forest was to be treated as an ecosystem, it no longer seemed sufficient to develop compartment models (taking into account physiological processes) and matter budget models because this would mask the heterogeneity of the interception of light and of the radiative climate. The complex sap-flow mechanisms related to the tree's hydric behaviour would also be overlooked. In the end, it would not be possible to follow the dynamics of the structural parameters, such as growth rings, in the trunks. Like their silviculture and agroforestry colleagues before them, the researchers developing the forest ecology approach therefore also recognized, albeit somewhat later, the need to switch to the scale of the individual tree.

Kurth initially became interested in AMAP's approach through Jean Dauzat's work on simulation of the radiative climate under plant cover. This work convinced Kurth to develop his own morphological model of individual trees, so as to be able to tackle them afterwards as a forest stand. He therefore agreed that it was necessary to take architecture into consideration. This was the key to the convergence in which he took part. In this regard, he often cited a forerunner in forest ecology itself; Adrian D. Bell, a British biologist and plant ecologist from the School of Plant Biology of the University College of North Wales. Bell was initially a specialist in rhizomes. Between 1972 and 1976, during his post-doctoral research at the Harvard Forest Centre in Massachusetts, he had taken advantage of the computer equipment available at the MIT Computing Centre in Amherst. He was intrigued by the regularity of the functioning of certain rhizomes. The angles of successive branches were often multiples of 60 degrees, and the general shape of the rhizome system therefore resembled a paving of opened or closed hexagons. This first led him to devise a simple hand-drawn graphical system of representation, in which the daughter branches grew in accordance with possibilities that had been calibrated in the field.[18] But since it was not possible to transcribe the dynamics of the rhizome, with its births and deaths, using these simple graphical marks, Bell came up with the idea of using a computer system that could illustrate the dynamics of growth and senescence over the course of time. He was convinced that it would be useful to retain a spatialized representation, and that it was necessary to use the computer in order to do so. Bell called his software RHIZOM, but he scarcely developed it, since he later began working in a less formalized manner on flowering plant morphology. At any rate, the common determination to spatialize (Bell) and to take architecture into account (de Reffye and Dauzat) played a key role for Kurth in the later convergence between logical models and architectural simulation in forest ecology. Thus, in 1994, Kurth developed his own software, christened GROGRA (for GROwth GRAmmar), on a Silicon Graphics IRIX 5.2 workstation. He kept Prusinkiewicz's L-system approach using turtles, but he considerably extended the rewriting rules so as to include more of the substance of the botanical laws that de Reffye and his team had highlighted earlier. To this end, Kurth met the AMAP researchers several times, and he recognized the debt he owed them. In order to move towards even greater botanical accuracy, Kurth was obliged to include no less than 26 different rules in his L-systems (as compared with Lindenmayer's original three). Since in this way the formalism became very intrinsically heterogeneous, *the priority no longer lay in emphasizing the theorems* that might, a priori, be drawn from it, even though this concern, which dates back to Lindenmayer, remained a consideration for Kurth. Kurth's main contribution was his idea of making the growth grammar sensitive to the environment right from the start. The purely local generativity of L-systems had played a large part in their success in a logicist and mechanistic context in developmental biology (Lindenmayer and Lück), and then in a constructivist context in computer graphics (Abelson, di Sessa and Prusinkiewicz). But in order to take account of the effect of the global control loops on the local and formal growth rules in an ecological modelling context, Kurth found himself obliged to temper this underlying mechanism in his own approach to morphogenesis. Certain entire branches might cast a shadow on the organs of other branches located lower down the tree, for example. In order to take account of this effect, which might be described as being ecological at the heart of the plant,[19] in the program the formal growth rules had to be able to constantly call on a representation of the tree's global architecture from both a topological and a geometric point of view. With this increased computing complexity – greatly assisted, it is true, by the use of object-oriented programming – Kurth was able to precisely calibrate GROGRA on a 14-year-old spruce by taking measurements similar to those of AMAP.

This shift towards architecture can also be seen in the forestry modelling work, starting in 1996, carried out by a team from the Finnish Institute of Forestry

Research at Helsinki. Inspired by AMAP, and more particularly by Kurth, Jari Perttunen's team chose to dispense with all logical or mathematical formalism from the start, and began straight away writing an object-oriented program in C++. The team called this program LIGNUM because it also was intended to consider the functioning rather than just the architecture. The model was purely computer-based. It considered the tree as "a collection of simple units corresponding to the organs".[20] In accordance with the approach imposed by C++, the tree was conceived as a collection of lists of basic units, using pointers to call up these units from each list. It was only at the level of the basic units' attributes that the morphology, the geometry, the carbon budget, etc. were implemented. There were only three types of basic unit: tree cross-sections; branching points; and buds. But this simplification was possible because, unlike AMAP's model, the model was not conceived to represent every type of plant, but only the Scots pine.[21]

In 1998 a publication helped to crystallize all these beginnings of convergences, and once again placed AMAP in the vanguard. This was the 46-page synoptic article co-authored by mathematician Christophe Godin and botanist Yves Caraglio, both from AMAP, which appeared in the *Journal of Theoretical Biology*.[22] After a recap of the requirements in botany by Caraglio, Godin gave an exhaustive, systematic and almost axiomatic presentation of his "multi-level model" concept that had been integrated into the AMAPmod program since 1996. The set of axioms for the graphs displayed an unprecedented flexibility. Leaving aside simulation, but making the most of the distinctions acquired by AMAP in its earlier simulations, this particular set of axioms emerges as a new type of formal modelling of language and of plant-structure analysis. It is possible to define the projection operators between the different scales of description within the plant. The very restrictive central hypothesis that Godin set for himself, however, was that the multiscale model had a recursive structure. In other words, the structure of the model did not depend on the scale on which the plant was described, even though on different scales certain different attributes emerged. As a result, a coherence constraint was imposed on the model.[23] This was what made the model similar to a mathematical model, yet without reducing it to a simple rule of fractal-type geometric self-similarity.[24] Certain theorems could be demonstrated. The emphasis placed on the axiomatic nature of the approach using graphs was initially due to Eric Elguero's prompting, but also and above all was inspired by Godin's early training and thesis work on speech analysis using stochastic models, including Markov chains. Already in his 1990 thesis, entitled "Proposal for a unified algorithmic framework for understanding continuous speech", Godin had proposed an "integrated architecture" of models so as to allow an intensive communication of the different standard models of speech recognition at various levels. Under the supervision of Bernard Dubuisson (University of Technology of Compiègne), he worked on producing algorithmic principles that would make it possible to establish connections between low-level numerical techniques and high-level symbol-recognition techniques. This unification was aimed at seeking a way to optimize integrated systems in the case of speech recognition. In 2000 Godin became a CIRAD AMAP researcher, temporarily assigned to INRIA to work in a team that had just been created, entitled "Virtual Plant". Over time, Godin's ties to AMAP gradually diminished and, several years later, although still working in closely connected fields, he became a full-time researcher at INRIA, and then director of the Virtual Plant team. Although an integrative approach (in the sense of multi-level) continued to permeate his work, his focus turned partly to the optimal algorithmic expression of the plant topology and geometry, partly to optimizing the digitization of plants, while at times focusing also on the numerical simulation of biophysical growth processes at the level of the meristems or of the architecture of certain organs rather than on the entire plant and on the fruit. Because of this, the integrative and explanatory botanical and biological approach is much less present in his work than in the both previous and the subsequent work of AMAP.

It should be noted that, at the turn of the 21st century, the ultimately classic argument of "the uniqueness of formalism for *understanding*" that was expressed so strongly in Godin and Caraglio's seminal article, was reflected in various other teams working on the subject. In Kurth, for example, it more explicitly took the form of a similar, but much more pragmatically oriented, argument in favour of "the uniqueness of formalism for *combining* different models" by means of L-systems.[25] Kurth recognized clearly, however, that his own enriched L-systems, like Godin's "multi-level model", barely still represented the mechanisms of phenomena, but that they could be used to integrate the different fragmented sub-models (topological, probabilistic, geometric, etc.) in the form of a meta-model that was somewhat more realistic than the initial logical models that emerged from the formal inflexibility of the first L-systems. We see thus that the inevitable dream of a theoretical and mono-formalistic unification, while perhaps rather premature, had been re-awoken, so to speak, at the very heart of the work of those who had initially been followers of AMAP and who were aware of an epistemology that was initially largely based on taking field measurements and their variability into account.

In 2002, however, this convergence – which, as we saw, verged on divergence (which in turn shows the power that AMAP had acquired in creating diverging research programmes within itself) – took on a different aspect. The convergence was brought about by a computer interface work carried out by Kurth himself, and by one of his students, Helge Dzierzon. They set GROGRA aside for a while in order to interface LIGNUM with AMAPmod. In this way, AMAPmod gradually became a standard of formal representation of plants in the more theoretical world of architectural modelling. As with some earlier cases in the history of computer science and software, there came a moment when compatibility was necessary in order for the dialogue to continue, and, as Dzierzon and Kurth wrote, "[the] intersubjectivity reduces the bias of our perception of ecological reality".[26] According to Dzierzon and Kurth, only in the presence of a common software, with a common formalization, would it be possible to continue to discuss ecological reality. However, a model structure, even when generalized, remained a simplification for the authors. The main role of interfacing was to expose the hidden assumptions in each model: the models could then be objectively compared and could criticize each other objectively, since this task had been delegated to the machine. For Kurth, who was pursuing his aim of finding the best formalism, the machine was of course a common ground. But it was a ground of *confrontation* between meta-models with the aim of selecting the most suitable, whereas for others – such as Dauzat – the machine remained a ground of *coexistence* between models for the purpose of carrying out empirical simulations. Through his interfaces, Kurth sought to make the meta-model distinct from the software so as to be able to establish its genericity.[27] This genericity itself was needed in order to *understand* and discuss what was happening in the models, independently of their computer implementation.

During the same period, INRA itself, after several years of hesitation, ended up spontaneously converging, from the inside, towards the techniques of architectural simulation. As a result, other approaches, which were competing with but also at the same time similar to the AMAP approach, emerged at the end of the 1990s, especially among the ecophysiologists.

Simulating the individual plant in order to observe crop functioning (1997–2000)

We saw earlier how the PIAF ecophysiologist Pierre Cruiziat had – more than others – become interested in the architectural approach focused on the individual plant, as proposed by AMAP. In his own way, he sought to have his laboratory adopt this approach. But he was not the only one with such interests at INRA. In 1995, under the supervision of Cruiziat's colleague Bruno Andrieu, from the INRA Bioclimatology Unit at Grignon, Christian Fournier began a thesis on "modelling interactions between plants in a population".[28] The initial idea of this work was that this need for plants to be dealt with at the level of the individual was no longer restricted to just forestry; it was also necessary for crop cultivation, because if the aim was to progress towards more environmentally friendly agricultural practices, then it would be necessary to accept that more complex systems were also involved in agriculture.[29]

Moreover, Andrieu had previously worked on systems to measure crop reflectance distances with the aim of relating these measurements to yield. He had rapidly recognized the limitations of this analytical and global approach: in the case of herbaceous plant crops, the directional properties of the leaves were not known. It was therefore impossible to determine a specific crop structure by working backwards from a global reflectance signal measured by sensors located at a distance (remote sensing): the measured signal could not be inversed. In the meantime, by 1991 Andrieu had learned of Dauzat's work on this issue. Together with Fournier, Andrieu therefore decided that he would also shift to a phase of architectural synthesis and simulation of the plant in order to apply this to the case of crop functioning. In Fournier's thesis, in particular, it was the maize's structure during its growth that was simulated. It is important to note that, for this simulation, they preferred to use the L-system formalisms of the *Virtual Laboratory*, which Prusinkiewicz's students had in the meantime made more permeable to different morphological traits. Having an L-system in the accessible form of a programming language made the implementation even easier. Furthermore, since the branching is not complex in the case of maize, it was not necessary to use the richest and most generalized of AMAP's simulation models. Fournier therefore used a parametric L-system in which the rules of production controlled the establishment and parallel evolution of a set of modules. These modules incorporated geometric and topological characteristics. They could replace each other or grow and co-evolve. Nonetheless, Andrieu, in collaboration with an INRA programmer, had to write a specific computer interface – just as Prusinkiewicz would do in 1996 – to take account of the effects of the global architecture on rules that would otherwise have retained purely local trigger conditions. Fournier and Andrieu were, in effect, faced with the same problem as Kurth had encountered in forest ecology. They decided not to reuse his L-system, however, which they considered too complex. With this ad hoc addition, Fournier managed to produce a visually realistic simulation of the morphogenesis of a maize plant by including phenomenological models of apex behaviour (with its leaf beginnings determined by a "law of response to temperature"[30]) and of leaf growth (phenomenological mathematical model of increase in length over time). The sub-models that they incorporated were thus highly phenomenological, and had a very limited basis in precise botanical or physiological knowledge.[31]

Thus it can be seen that the basis of Fournier and Andrieu's work is their reuse of the idea of multi-modelling[32] or pluriformalization in order to apply it to the case of crops. As in AMAP's case, it was the computer infrastructure of the L-systems, together with their step-by-step processing of the complex phenomenon of morphogenesis, that enabled them to do this. Thus, in the ecophysiology of crops there was also a clear convergence of certain works towards the philosophy that AMAP, for its part, had been advocating since the mid-1980s.

The association between AMAP and INRIA: sub-structures and factorization (1998–2006)

Nevertheless, this evolution towards a strong, software-based and thus almost technological convergence was not the only motivation behind architectural modelling at the end of the 1990s. De Reffye was already no longer participating directly in this convergence, which, as we have seen, was taking place essentially on a software-based and formal level, and was moving further away from field-work and becoming more theoretical once again. The reason for this distancing was because, between 1998 and 2002, spurred by the momentum created by AMAPhydro and by de Reffye's correlated renewed concern for concrete agronomic issues, de Reffye was considering his own return to mathematics via the control theory. He was soon given an opportunity: in 1996, during the external review of AMAP, an INRIA researcher, Olivier Monga, had been one of the reviewers. He had been favourably impressed by AMAP's work and he had remained in contact with de Reffye since then. De Reffye had even selected Monga as permanent scientific advisor. On 27 January 1997, however, following an agreement with the Chinese Academy of Science, INRIA created LIAMA, the Franco-Chinese Laboratory of Informatics, Automation and Applied Mathematics,[33] and Olivier Monga became its first French co-director. The LIAMA headquarters were set up in the Beijing Institute of Automation. It was established as a long-term cooperation structure to foster the realization of targeted projects.[34] Monga suggested to de Reffye that he should join him in Beijing. There was a community of interests, since de Reffye was also seeking to develop a more powerful AMAP software platform. At the beginning of 1998 CIRAD therefore temporally assigned de Reffye to INRIA and, initially accompanied by Frédéric Blaise, he joined Monga at LIAMA to develop their project on "Stochastic, functional and interactive computer modelling of plant growth". Their Beijing partners were specialists in automation, as well as agronomists from the Chinese University of Agronomy. De Reffye considered their approach to be more practical and less biased on the subject of modelling and simulation. He had thus found the freedom he had hoped for in leaving his position as director. Over the course of three years, he managed to supervise eight doctoral students. The main result was the discovery of a method of simplifying the formal representation of plants by using stochastic simulations. The benefit of this simplification was that it made it possible to considerably increase the speed of calculation and to integrate the plant's physiological behaviour in an acceptable way.

Initially, de Reffye bore in mind that AMAP's founding principle was to trust in the stochastic simulation program and allow it to interpolate between situations that had actually been observed in trees in the field. While working with the automation specialists from INRIA and the Chinese Institute of Automation, however, de Reffye became convinced that this interpolation led to a loss of control over what the program was doing. Simulation replaced modelling in the program, and it was here that the most numerous and time-consuming calculations had to be made. The LIAMA team therefore tried to systematize de Reffye's habit of seeking cases that could be calculated by hand or at the very least through mathematical equations. It should be recalled that the aim of this practice was initially to verify the simulation. In this way, the team rather unexpectedly managed to demonstrate that nearly 90 per cent of cases of simulated trees could in fact be calculated in this mathematical – and therefore calculation-saving – manner. De Reffye and his students, including Xin Zhao and Hong-Pin Yan, discovered that this was because the meristems often ultimately generate only a small number of typical sub-structures in order to create the entire tree. Some simulated trees might contain the same sub-structure as many as 600 times. The 1991 "reference axis" was therefore abandoned in the new software, which was christened GreenLab for this occasion. In 2001, initially inspired by Godin's approach of nested sets, but having modified it in accordance with their observations regarding both the functioning of stochastic simulations and the observed real-life hierarchy of botanical architectural units, Xin Zhao and de Reffye proposed the notion of a "dual-scale automaton model". This was a two-level automaton; an automaton of automata. Within one macro-state, micro-states can move towards each other using fixed laws of probability.

The notion of physiological age played a pivotal role in this revival of morphological hierarchization because the sub-structures were considered to be equivalent and required no further calculations if they were of the same physiological age. This meant that the sub-structures all had the same set of hidden parameters. De Reffye proposed that four levels or scales should be identified: the metamer level (composed of an internode and its axillary leaves, fruits and buds); the growth-unit level (the set of all the metamers that appear during the same growth cycle); the level of the load-bearing axis; and the level of the "sub-structure" itself, namely the branch, in the case of plants with branches.[35] In order to construct a simulated tree, it was no longer necessary to rely on the scale of the metamers from the outset. A sort of sub-structure was calculated once and for all, and once the automaton had ordered its reiteration with a certain probability, all the program had to do was to retrieve from the memory the content of the pre-calculated topological and geometric parameters in order to display this new sub-structure.[36] Almost no further calculation was then necessary. As a comparison, a 20-year old cherry tree took two minutes to be calculated by AMAPsim, and less than one second using GreenLab. In this specific case, the architectural simulation was therefore approximately 200 times faster.[37] This new computer model of course still used a Monte-Carlo type simulation, but in a much more limited manner than previously. In order to achieve the main aim – the coupling with ecophysiology in order to create a structural-functional model – GreenLab largely reused the concepts and equations of AMAPhydro by considering the production of biomass, leaf by leaf, as a function of the architecture.[38] Finally, in this regard, it should be noted that because of this remarkable reduction in the number of computation steps necessary, GreenLab functioned regularly from 2002 on a simple PC running the Matlab formal calculation software.[39] This conceptual progress, together with the technological advances that have occurred in the micro-computing industry since then, have meant that AMAP could finally give up the systematic need for very rare and expensive graphics workstations. In turn, this has facilitated its international distribution as well as making it easier to compare its performance with that of other models.

The structural-functional model was then rapidly calibrated on plants – initially on plants without branches, such as the sunflower, cotton plant or maize. The collaboration of the Chinese University of Agronomy was invaluable here: it was they who dealt with supplying the necessary experimental data for the calibration. The calibration technique was made easier thanks to the formal simplifications due to the AMAPhydro step and to the development of the dual-scale automaton. This was one of the immediate effects of *remathematizing* the AMAP model: relatively classic statistical *fitting* procedures such as generalized least squares could finally be applied directly in order to economically identify the parameters.[40] Everyday use of GreenLab in agronomy thus finally seemed feasible.

Another major outcome of this remathematization began to appear after 2001. The remathematization was implemented by de Reffye at INRIA in Rocquencourt, where he had been temporarily assigned after his return to France in 2002. This outcome was the possibility of determining the optimal crop management beforehand by using the calculation techniques arising from the optimal control theory and automation. In the context of the METALAU[41] project, with the assistance of Maurice Goursat and Jean-Pierre Quadrat from INRIA, Boa-Gang Hu, Paul-Henri Cournède (from INRIA and the École Centrale Paris) and Philippe de Reffye had recently explained the dynamic equations of the GreenLab model. These were

Figure 7.1 Tree simulated in 2006 by the Digiplante software from the École Centrale Paris (GreenLab team), taking into account the growth/development retroaction. Copy taken from the thesis of Amélie Mathieu, with her kind authorization.

essentially recurrence equations. The automata could be translated in terms of matrix equations. The computer formalism was thus *remathematized* insofar as it was replaced by an algebraic formalism in which we find matrix multiplications (two-dimensional tables).[42] Both Quadrat, and especially Goursat, had participated in the conception of the Scilab software for the numerical resolution of optimization problems in science. This platform, developed in partnership with the École des Ponts[43] since 1990, has been a free software since 1994. The authors intended to integrate GreenLab in Scilab. This project retrospectively and publicly justifies the use of the "-Lab" suffix.

The return to equations thus made it possible to "de-spatialize" to a certain extent the formalism of morphogenesis in order to reinstate in it the linguistic linearity of the algebraic equation or, at the very least, of the combinatorial analysis. The equation itself, while it cannot be inversed in order to give the optimal values for a given plant production directly, can at the very least be manipulated by fairly classic numerical optimization techniques. In the end, the plant would thus resemble an artificial dynamic system whose optimality would be more controllable a priori. The simulation remathematization work carried out by the Digiplante laboratory (a joint CIRAD/INRIA team led by de Reffye after his return from China) was thus based on the complexity of the AMAP simulations, but proposed a "structural factorization of the plant"[44] that played a part in relinquishing the simulation solution. This work thus tended to combine architectural simulations with the process-based models resulting from ecophysiological modelling.

In the meantime, the institutional convergence was accelerating rapidly. On 1 January 1999, on the expiry of the 1995 partnership agreement, and spurred on by the policies of the French Minister for Research of the time, Claude Allègre, AMAP became a Joint INRA/CIRAD Research Unit. Starting in 1998, what was still sometimes seen as an expensive offshoot in the context of CIRAD's scientific policy became, on the contrary, a showcase for the reorganization of French research around "centres of expertise". The methodology, concepts and software technologies of AMAP were also taught more widely and systematically, in particular at the École Centrale Paris, from 2002 onwards. This undertaking required the participation of all the researchers. On 1 January 2001, under the leadership of François Houllier, the convergence continued: AMAP became a UMR – a Joint Research Centre – comprising CIRAD, INRA, CNRS and the Montpellier II University, under the name "Joint Botanical and Bioinformatics Research Centre for Plant Architecture". The use of the term "bioinformatics" was a hard-won victory, in particular for François Houllier, over the monopoly of genomics and proteomics. In January 2003 Houllier was named Head of the INRA Department of "Forest, Prairie and Aquatic Environments" (EFPA – *Écologie des forêts, prairies et milieux aquatiques*), while Daniel Barthélémy became Director of the AMAP Joint Research Centre, UMR AMAP 5120. Also on 1 January 2003, this Joint Research Centre joined forces with EPHE (the "École Pratique des Hautes Études"), INRIA and IRD.

Recap: pluriformalized simulation and convergence between disciplines

This third age of models, which I propose to call the age of convergences, certainly deserves its name. After a period of disseminating the formalized proposals of mathematical modelling and then of speculative computer simulation,[45] the 1970s witnessed the emergence of the first convergences between simulations and the empirical. This first series of convergences was limited. In the case of biologists and botanists, it fell under the category of concerns that were still primarily theoretical. Pragmatic mathematical modelling – the source of which resided in British biometry and its related epistemology – was unaffected by such developments. It was a different case, as we have seen, once this convergence between simulation and the empirical became a necessity in agronomical and field-related problems. In this context, simulation immediately entered into close dialogue with pragmatic and mono-formalized statistical modelling. In the history I have recounted, we saw that the computer, which very early on was considered by some to be a support for realistic, detailed and pluriformalized simulations, was able to fight every step of the way against this more classic type of modelling, and ultimately took the crown – perhaps ephemeral but nonetheless real – even in the case of explanatory modelling of morphogenesis.

Thus began the era of the three disciplinary convergences or conciliations: with computer graphics; with forestry; and with automation and discrete mathematics. This series of convergences went hand in hand with a formal simplification, from the point of view of the mathematical generality of the models supporting the simulations. This simplification led to a standardization of the software tool, as well as to an expansion of its operational nature. But this remathematization would doubtlessly not have been possible if the spread of pragmatic models had not first been exorcized or neutralized by the inherent pluriformalism of architectural simulations. As a consequence of this, and despite considerable resistance, we have seen that scientists had to urgently reform their epistemology itself in order to be able to consider something that had become very real and that functioned before their very eyes. Today, some work has gone beyond even this level, and proposed what could be called a fourth convergence: the convergence of architectural and morphological simulation with genetics and molecular biology.[46]

At the same time as this development of architectural simulation, proposals for physicalist modelling continued to emerge more or less everywhere throughout the world, and to occasionally seek calibration, but only on specific plant species. Indeed, these proposals remained largely theoretical and speculative, which is why I have not reported in detail on their evolution over the last three decades: they have merely contextualized from a distance the advances of operational architectural simulation, without fundamentally renewing the spirit of the monoformalized approaches of the second period: the period of dispersions. These proposals remained more or less at that stage of the history of models. They did not converge towards each other, nor did they converge towards specific uses. This was the case, for example, of the biomechanics work of the botanist and mathematician Karl J. Niklas, of the Plant Biology Department of Cornell University in New York. From a joint mechanical engineering and evolutionary biology perspective, Niklas used the computer to complexify Murray and Rashevsky's approach. He proposed to use the equations of fluid mechanics in order to render temporal on a phylogenic level a principle of physical and physiological optimality regarding what he called the plant "specifications". Niklas and his colleagues thus ended up producing a program that simulated the natural selection over time of the physical and physiological specifications of a large number of plants whose shapes had first been selected at random.[47] They hoped the most efficient forms – from the point of view of their allometry – would emerge.

Two Israeli researchers, Tsvi Sachs and Ariel Novoplansky (from the Botany Department of the Hebrew University of Jerusalem), the colleagues and successors of Dan Cohen, also emphasized the evolutionary nature of plant morphology, but on another scale. According to them, "architectural models do not suffice"[48] to express the architecture of a plant. They therefore did not endorse Hallé and Oldeman's approach, which they considered to be reductive. Indeed, during the growth of the plant – which they pointed out belonged to a late evolutionary group – unexpected details appeared. These details were constrained by the mechanisms involved during growth, and by progressive differentiations that are in turn a function of the distance from the apices to the roots: the plant remains an evolutionary organism even on an ontogenetic level. The authors therefore argued that the mechanisms that are responsible for the gradual change in the mode of development of the branches were not known:[49] it would therefore be wrong to represent them using a deterministic model and leave it at that. In fact, in 1995, when invoking the lack of explanation in the work of colleagues, Sachs and Novoplansky were referring in particular to the school of Prusinkiewicz,[50] since at that time the latter had not yet integrated into their models the unforeseen events or the feedback of the global on the local, as Kurth did later by learning from AMAP. Furthermore, they recognized the more biochemical and mechanical level of their morphogenetic work. In 2004, in connection with a local self-organization model, Sachs and his colleagues demonstrated the role of auxins (a class of plant growth hormones) in the formation of leaves. Since then, the work of Sachs and Novoplansky has continued to highlight the plasticity of the morphogenetic mechanisms in plants, in particular in response to various interactions with the environment.[51]

During the 1980s and 1990s, new, purely physicalist models also regularly appeared, in pace with advances in physical chemistry. This was the case for the analogical model of diffusion-limited aggregation (DLA) produced by T.A. Witten and L. Sander (University of Michigan, 1981), experts in statistical physics far from equilibrium. This model was created in the context of studies on the phenomena of crystal growth and percolation. It can be easily simulated by computer and results in fractal growth. The aim was to produce, by a random process, a particle that would spread over a pre-constituted aggregate, and to incorporate this particle as soon as it came into contact with the aggregate. The growth of a branching form is thus obtained and has a purely random and destructured nature that cannot, of course, be closely or precisely compared with plant growth. Indeed, plants do not grow by the aggregation of particles floating in the external aerial environment, but by internal growth and preliminary assimilation of nutrients. This model was therefore of no interest to botanists.

Furthermore, starting in 1991, S. Douady and Y. Couder, from the Laboratory of Statistical Physics at the École Normale Supérieure campus (on rue Lhomond), described an experiment on self-organization of drops of ferrofluid in the presence of a magnetic field: starting from a small platform in the oil into which they are added, these drops – periodically released from a pipette – spread towards the area of least energy depending on their polarization, since they are dipoles, and depending on the overall magnetic field. Thus each new drop settles at a given distance from the previous drop, in such a way that the classic irrational divergences of phyllotaxis (Fibonacci series) are reproduced.[52] The related digital simulation produced the same results. This was a new occasion for the Canadian bio-mathematician Roger Jean to declare, in true d'Arcy Thompson manner, that there was no need to imagine that the *patterns* of phyllotaxis were controlled by genes, and that it was enough to link them to physical laws.[53] Jean was still obliged to recognize, however, the immense variety of explanatory approaches in phyllotaxis, which represents just one tiny part of morphogenesis.

None of these monoformalized models has truly reached the level of the empirical in its precision, or in its generality and diversity. The various analogical substrates of physicalism have at the most become more diversified, without putting an end to the display of scattering between all these disjointed approaches, which are often purely evocative from a theoretical point of view. Despite some quivers of convergence between the classically theoretical and physicalist approaches, computer simulation of the whole plant, which is conceived on the basis of the actual knowledge of the descriptive science of botany, still seems likely today to retain the lead for some time yet in the domain of transferring formal methods to the field. Furthermore, the search for mathematical models based on architecturally faithful simulations with the same value as a virtual field of experimentation – rather than the search based on the usual physicalist suggestions inspired directly by chemistry or physics, or on the new "computationalist" or "algorithmicist" suggestions that were intended to reignite the dream of direct formal unification by means of meta-models – blossomed and took on an undeniable epistemological meaning.

Indeed, as we have seen, this type of undertaking in architectural simulation, paired with a modelling on simulation, basically owes its value to the fact that it is based on a considerable body of previous field work, although without stemming from the direct and classic dialectic relationship between an actual experience and a theoretical formalism: it is here that another term enters the scene, that of simulation, where pluriformalization and the sub-symbolic use of symbols (simulation as a partially expanded formal representation) are fundamental. Such a simulation is thus no longer comparable with a purely positivist mathematization, which is carried out "by intuition" or according to a preference for a given formalism (theories of catastrophe, fractals, categories, graphs, automata, L-systems, etc.) and using field data in an arbitrary and undifferentiated way. Since such a simulation is a process of formalization that is neither dialectic nor positivistic, the method that it introduces is a rather unusual manner of testing and then seeking models. This way of simulating first via the computer implementation of complex integrative simulation models opens the way to a new and more efficient search for mathematical models. This is because such a search is based on serious and knowledge-laden simulations, and not directly on data or on purely phenomenological models of data.

Notes

1 The École Centrale Paris, also known as CentraleSupelec since 2015, is a long-established and illustrious French institute of research and higher education in engineering sciences.
2 Pierre Cruiziat is an Agricultural Engineer from Paris-Grignon (now called the "Agro-ParisTech"). In the early 1970s he obtained his Doctorate in Plant Physiology. His research work at INRA dealt essentially with water transfer in plants. In 2003 he worked in tree ecophysiology at the Clermont-Ferrand branch of INRA. Since 1990

he has been attached to the PIAF unit – i.e., *Physique et Physiologie Intégratives de l'Arbre en environnement Fluctuant* [Integrative physics and physiology of trees in fluctuating environments] – which has been affiliated with the Blaise Pascal University since that date. Cruiziat has been fighting for the spatialization of functional models for two reasons: 1) the necessary consideration of the *geometric heterogeneity* of the plant canopy, and thus of insolation; and 2) the necessary consideration of the *morphological heterogeneity* of the organs in the tree-crown and the internal competitions within it.

3 See Reffye (de) (P.), Blaise (F.), Fourcaud (T.), Houllier (F.), Barthélémy (D.), "Un modèle écophysiologique de la croissance et de l'architecture des arbres et de leurs interactions" [An ecophysiological model of the growth and architecture of trees and their interactions], Andrieu (B.) (Ed.), *Modélisation architecturale [Architectural modelling]*, Proceedings of the seminar of 10–12 March 1997, Bioclimatology Department, Paris, INRA-Grignon, 1997, p. 129–135.

4 Ibid., p. 130.

5 For both the equation and its legend, see ibid., p. 131.

6 Reffye (de) (P.), Blaise (F.), Chemouny (S.), Jaffuel (S.), Fourcaud (T.), "Calibration of a hydraulic architecture-based growth model of cotton plants", *Agronomie*, 1999, Vol. 19, No. 3/4, pp. 265–280.

7 Malézieux (E.), Trébuil (G.), Jaeger (M.) (Eds), *Modélisation des agroécosystèmes et aide à la décision* [Modelling of agro-ecosystems and decision-support tool], Joint edition: Montpellier and Versailles, CIRAD and INRA Bookstore editions, 2001, p. 160.

8 Szilard (A.L.), Quinton (R.E.), "An interpretation for *DOL* systems by computer graphics", *The Science Terrapin*, 1979, Vol. 4, pp. 8–13.

9 See Abelson (H.), "Logo graphics as a mathematical environment", *Proceedings of the Annual Conference of the Association of Computing Machinery, ACM-CSC-ER*, Houston, ACM-Press, 1976, p. 160: "All mathematics has (ultimately) intuitive foundations, but in turtle geometry the links are particularly close".

10 Ibid., p. 159.

11 Meristem (see Glossary), located at the tip (apex) of a growing branch or shoot.

12 For a history of the "Artificial Life" programme, see Goujon (P.), "La biologie à l'ère de l'informatique. Connaissance et naissance de la vie artificielle, première partie" [Biology in the computer age. Knowledge and origin of artificial life, part one], *Revue des Questions Scientifiques* [Journal of Scientific Questions], 1994, Vol. 165, No. 1, pp. 53–84; Goujon (P.), "La biologie à l'ère de l'informatique. Connaissance et naissance de la vie artificielle, seconde partie" [Biology in the computer age. Knowledge and origin of artificial life, part two], *Revue des Questions Scientifiques* [Journal of Scientific Questions], 1994, Vol. 165, No. 2, pp. 119–153.

13 Prusinkiewicz (P.), Lindenmayer (A.), *The Algorithmic Beauty of Plants*, New York, Springer Verlag, 1990; reprint: 1996, p. 193.

14 Prusinkiewicz (P.), "Modeling plant growth and development", *Current Opinion in Plant Biology*, 2003, Vol. 7, No. 1, p. 2.

15 What de Reffye referred to here as "laws of nature" are, for example, the architectural models, the stochastic nature of meristem behaviour in botany, the reference axis and physiological age of the meristems, or the water efficiency law in agronomy. In other words, they are phenomenological laws on a fairly global scale. De Reffye compared them with Ohm's law in electricity ($V = RI$), and considered that it would take a new Maxwell to remove their purely phenomenological nature and deduce them from explanatory models. According to de Reffye, these laws would not evolve with regard to their substance, but only with regard to their form, in their formalism, which, for that matter, was something to be desired. This is why they are laws of *nature*.

16 Prusinkiewicz (P.), "Modeling of spatial structure and development of plants: a review", *Scientia Horticulturae*, 1998, Vol. 74, Nos 1–2, p. 114.

17 Kurth (W.), "Stochastic sensitive growth grammars: a basis for morphological models of tree growth", proceedings of the Symposium on *L'arbre: Biologie et Développement* [The Tree: Biology and Development], Montpellier, 11–16 September 1995, p. 12.

18 Even though Bell's system was simpler due to its more limited range of subject matter (i.e., the rhizome), we can see that Bell was proposing a very similar stochastic simulation solution to de Reffye's, in a similar time-frame. In his 1976 article, Bell appreciated having been able to use something other than plotters, namely one of MIT's first graphics screens; this enabled him to dynamically simulate, within a single image, the loss through rot of the rhizome parts affected by senescence. Cf. Bell (A.D.), "Computerized vegetative mobility in rhizomatous plants", in A. Lindenmayer, G. Rozenberg (Eds), *Automata, Languages, Development*, Amsterdam, North-Holland Publishing Company, 1976, p. 5. As we know, it was not until 1989, i.e., 13 years later, that de Reffye was able to benefit from equivalent graphics technologies.

19 Like AMAP before him, for that matter, Kurth invoked James White's seminal article: "The plant as a metapopulation", *Annual Review of Ecology and Systematics*, 1979, Vol. 10, 109–145.

20 Perttunen (J.), Sievänen (R.), Nikinmaa (E.), Salminen (H.), Saarenmaa (H.), Väkeva (J.), "LIGNUM: a tree model based on simple structural units", *Annals of Botany*, 1996, Vol. 77, No. 1, p. 87.

21 Ibid., p. 96.

22 Godin (C.), Caraglio (Y.), "A multiscale model of plant topological structures", *Journal of Theoretical Biology*, 1998, Vol. 191, No. 1, 1–46.

23 Ibid., p. 37.

24 Godin's more recent work (2005), especially that published in conjunction with Prusinkiewicz, shows however that the aim of highlighting a sort of general self-similarity has not been abandoned.

25 This shift is very clear in his latest publications. He is trying to interface many other models so as to be able to compare them on an objective basis, i.e., from a quantifiable point of view. See Kurth (W.), "Spatial structure, sensitivity and communication in rule-based models", in Franz Hölker (Ed.), *Scales, Hierarchies and Emergent Properties in Ecological Models*, Frankfurt am Main, Peter Lang, 2002, pp. 95–104; Dzierzon (H.), Kurth (W.), "LIGNUM: A Finnish tree growth model and its interface to the French AMAPmod database", in Franz Hölker (Ed.), ibid., pp. 29–46.

26 Ibid., p. 46.

27 Kurth (W.), 2002, art. cit., p. 99.

28 See Fournier (C.), "Modélisation des interactions entre plantes au sein des peuplements. Application à la simulation des régulations de la morphogenèse aérienne du maïs (Zea mays L.) par la compétition pour la lumière" [Modelling of plant interactions within stands. Application to the simulation of the rules of aerial morphogenesis of maize (Zea Mays L.) by competition for light], INAPG thesis, April 2000, p. 1.

29 This argument also stems from the fact that the thesis was financed by the "Development of fuel-efficient and clean agricultural processes" programme of the French environment and energy management agency (ADEME).

30 This law is found in ecophysiological literature from the early 1980s: see Fournier (C.), Andrieu (B.), "Utilisation de l'approche L-système pour la modélisation architecturale du développement du maïs" [Use of the L-system approach for the architectural modelling of maize development], in B. Andrieu (Ed.), 1997, *op. cit.*, p. 205.

31 De Reffye thus claims that, on the contrary, his own sub-models all reflect authentic elementary plant characteristics, rather like "laws of nature". In this regard, he saw a considerable difference at that time between the work of Andrieu's team and his own. Nonetheless, this notion of "law of nature" is no doubt very questionable: on what level should it be taken? We have seen that de Reffye was haunted by this issue from the moment in 1974 when he decided that the probabilities of coffee-plant flowering were

"genetic traits" in their own right. Were the phenomenological and local (at the level of the organ) laws of ecophysiology nonetheless so different in nature from the laws of random meristem branching? As can be seen, the problems (and the corresponding fear) of detachment are once again taken up on this more detailed level. De Reffye was in no way a supporter of the fictional in modelling, as we can ultimately see.

32 See Glossary. The term "multi-modelling" was introduced in 1989 by the Turkish-born Canadian computer scientist Tuncer Ibrahim Ören, in the context of a reflection on the paradigms of computer simulation aimed at essentially technological and industrial systems. The term was intended to indicate a simulation infrastructure capable of combining different types of modelling, including, at the start, models with differential equations and discrete models. It was adopted and further developed, along with the notion of "multiformalism", in 1991 and 1993 by simulation specialists such as Herbert Praehofer, and later Bernard P. Zeigler. See Zeigler (B.P.), Praehofer (H.), Kim (T.G.), *Theory of Modeling and Simulation: Integrating Discrete Event and Continuous Complex Dynamic Systems*, New York, Academic Press, 1976; reprint 2000, pp. 227–229. Since that time, the concept has been applied more widely in ecology and sociology, where individual-centred simulation had already emerged at the end of the 1980s. Regarding multi-modelling in ecology, see Duboz (R.), "Intégration de modèles hétérogènes pour la modélisation et la simulation de systèmes complexes – a pplication à la modélisation multi-échelles en écologie marine" [Integration of heterogeneous models for the modelling and simulation of complex systems – application to multi-scale modelling in marine ecology], Computer science thesis, Littoral Côte d'Opal University, 2004, pp. 16–17. Regarding individual-centred simulation in ecology, see Grimm (V.), "Ten years of individual-based modeling in ecology: what have we learned and what could we learn in the future?", *Ecological Modelling*, 1999, Vol. 115, 129–148; in sociology, see Amblard (F.), "Comprendre le fonctionnement de simulations sociales individus-centrées – application à des modèles de dynamiques d'opinion" [Understanding the functioning of individual-centred social simulations – application to models of opinion dynamics], Computer Science thesis, Blaise Pascal University – Clermont II, 2003.

33 Now called the "Sino-European Laboratory of Informatics, Automation and Applied Mathematics".

34 This international "joint laboratory" type of research structure was not new. It had been an important approach in CNRS policy since the 1960s. With regard to INRIA, which was set up in 1967 in the context of the implementation of the "Calculation Plan", the most specific goal was to discover potential uses for industry. Article 2 of Decree 85-831 of 2 August 1985 specifies moreover that one of INRIA's missions is "to organize international scientific exchanges".

35 Reffye (de) (P.), Goursat (M.), Quadrat (J.P.), Hu (B.G.), "The dynamic equations of the tree morphogenesis GreenLab Model", in B.G. Hu, M. Jaeger (Eds), *Plant Growth Modeling and Applications, Proceedings of the 2003 International Symposium on Plant Growth Modeling, Simulation, Visualization and their Applications (PMA03)*, Beijing, China, 2003, p. 109.

36 The probability of the sub-structure can be constructed through integral stochastic simulation. It is thus possible to empirically compare the variability simulated by AMAPsim and the variability modelled and simulated by the mixed model with stochastic sub-structures. Cf. Kang (M.Z.), Reffye (de) (P.), Barczi (J.F.), Hu (B.G.), "Fast Algorithm for Stochastic Tree Computation", 11th International Conference in Central Europe on Computer Graphics – Visualization and Computer Vision 2003, in cooperation with *EUROGRAPHICS, Journal of WSCG (Winter School of Computer Graphics)*, Vol. 11, No. 1, p. 5: the realism of the variability of the two different types of images produced is indistinguishable to the naked eye.

37 Ibid., p. 4.

38 Yan (H.P.), Reffye (de) (P.), Le Roux (J.), Hu (B.G.), "Study of plant growth behaviors simulated by the functional-structural plant model GreenLab", in B.G Hu, M. Jaeger (Eds), op. cit., pp. 118–122.

39 With a 1 700 MHz processor and 256 Mbytes of memory. Cf. Kang (M.Z.), Reffye (de) (P.), Barczi (J.F.), Hu (B.G.), art. cit., p. 4.

40 Zhan (Z.G.), Reffye (de) (P.), Houllier (F.), Hu (B.G.), "Fitting a structural-functional model with plant architectural data", in B.G Hu, M. Jaeger (Eds), op. cit., p. 241.

41 The acronym signifies "*METhode, Algorithmes et Logiciels pour l'AUtomatique*" (method, algorithms and software for automation). This project was led by Maurice Goursat, Research Director at INRIA, who took over from Jean-Pierre Quadrat, Research Director at INRIA, and ex-supervisor from 1987 to 1999 of a similar project entitled META2. Both Goursat and Quadrat were specialists in optimization methods, in particular using stochastic approaches.

42 Reffye (de) (P.), Goursat (M.), Quadrat (J.P.), Hu (B.G.), op.cit., pp. 109–111.

43 Like the École Centrale Paris, the École des Ponts is a famous and long-established French institute of research and higher education in engineering sciences.

44 See Cournède (P.H.), Kang (M.Z.), Mathieu (A.), Barczi (J.F.), Yan (H.P.), Hu (B.G.), Reffye (de) (P.), "Structural factorization of plants to compute their functional and architectural growth", *Simulation*, 2006, Vol. 82, No. 7, pp. 427–438.

45 The two previous periods – of the dissemination of mathematical modelling, and then of the first speculative computer simulations of biological growth and morphogenesis – have been described and analysed in another book: Varenne (F.), *Formaliser le vivant* [Formalizing living beings], Paris, Hermann, 2010.

46 The same holds true for the work on QTL (*Quantitative Trait Loci*). These are the *loci* of genes that code for quantitative traits, and in particular for the sizes of morphological traits. In 2000, German researchers at the Gatersleben Plant Genetics Institute performed computer simulations of the end morphology of a spike of barley in connection with known genotypes by using QTLs. They simulated only one part of the herbaceous plant, but already "the interactions between genes and between alleles are calculated using the knowledge furnished by experimental genetic studies [and] allowing the values of morphological variables to be predicted", Buck-Sorlin (G.H.), Bachmann (K.), "Simulating the morphology of barley spike phenotypes using genotype information", *Agronomie*, 2000, Vol. 20, 691. These predictions were made by interpolation. The authors used the formalism of Prusinkiewicz's "virtual laboratory". They used a parametric L-system similar to that of Fournier and Andrieu. Since 2000, several teams have continued to develop this fourth convergence. See, for example, the work based on Kurth and Buck-Sorlin: Xu (L.), Henke (M.), Zhu (J.), Kurth (W.), Buck-Sorlin (G.H.), "A functional-structural model of rice linking quantitative genetic information with morphological development and physiological processes", *Annals of Botany*, 2011, Vol. 107, No. 5, 817–828. Similar work has been based on de Reffye and Cournède: Letort (V.), Mahe (P.), Cournede (P.H.), de Reffye (P.), Courtois (B.), "Quantitative genetics and functional-structural plant growth models: simulation of quantitative trait loci detection for model parameters and application to potential yield optimization", *Annals of Botany*, 2008, Vol. 101, No. 8, 1243–1254.

47 See Niklas (K.J.), "Computer simulations of branching-patterns and their implications on the evolution of plants", in L.J. Gross, R.M. Miura (Eds) *Some Mathematical Questions in Biology, Lecture on Mathematics in the Life Sciences*, 1986, Vol. 18, American Mathematical Society, Providence, Rhode Island, 1–50; Niklas (K.J.), "The evolution of plant body plans: a biomechanical perspective", *Annals of Botany*, 2000, Vol. 85, No. 4, 411–438; Niklas (K.J.), Spatz (H.C.), "Growth and hydraulic (not mechanical) constraints govern the scaling of tree height and mass", *PNAS*, 2004, Vol. 101, No. 44, 15661–15663.

48 See Sachs (T.), Novoplansky (A.), "Tree form: architectural models do not suffice", *Israel Journal of Plant Sciences*, 1995, Vol. 43, No. 3, 203–212.
49 Ibid., p. 209.
50 At the time, they were unaware of AMAP's approach, and in particular of the work of Frédéric Blaise.
51 See, in particular, Sachs (T.), "Consequences of the inherent developmental plasticity of organ and tissue relations", *Evolutionary Ecology* 2002, Vol. 16, 243–265; Gruntman (M.), Novoplanksy (A.), "Ontogenetic contingency of tolerance mechanisms in response to apical damage", *Annals of Botany*, 2011, Vol. 108, No. 5, 965–973.
52 See Douady (S.), Couder (Y.), "Phyllotaxis as a physical self-organized growth process", *Physical Review Letters*, 1992, Vol. 68, No. 13, 2098–2101; Jean (R.V.), *Phyllotaxis: A Systemic Study in Plant Morphogenesis*, Cambridge, Cambridge University Press, 1994; reprint: 1995, pp. 262–265.
53 Ibid., p. 264.

8 Twenty-one functions of models and three types of simulations – classifications and applications

Up to this point, this book has presented a comparative, historical and epistemological analysis of the different types of models and digital simulations used in the study of plant morphogenesis. It has illustrated the wide range of approaches, along with the rivalry that exists between them. It has also shown that, despite these competing approaches, a historical unifying tendency has clearly emerged: the growing importance of integrative simulations that are, in turn, backed by more complex simulation models. In order to better explain what is at stake in these unprecedented aspects of the computerization of science and technoscience, this new chapter aims first of all to gain a better overview of the issue by first examining it from the level of philosophical and conceptual analysis, and by then returning to review the specific cases that were presented earlier.

The characterizations and classifications that I suggest in this new chapter are the result of epistemological work carried out since 1999 based on a series of historical investigations on models and simulations in biology, sociology and geography. These classifications have already been published in part, but largely in French and in publications not directly related to my work in the field of history of science. The publication of this updated English translation of my 2007 work gives me an opportunity to combine these two types of studies so as to offer a results-based assessment of the usefulness of these epistemological classifications in one single work.

In the first section of this chapter, I will show that it is necessary to differentiate between three characteristics of models that are often confused in the literature, namely the model's epistemic function, its substantial nature, and its functional principle. I will then identify three levels of epistemic functions of a model: the general function, the main function and the specific function. This series of distinctions will lead to the identification of twenty-one specific functions of models. Having listed these, I will then expand on certain particular points concerning the different natures and principles of models. Thereafter, I will present a general characterization of simulations, which will make it possible to clearly distinguish between simulations and models. This characterization will be followed by a classification for distinguishing the three different types of computer simulation. Lastly, with the aim of demonstrating the relevance and usefulness of all these characterizations, I will apply them to several models and simulations that I have come across in the course of my longitudinal case study regarding

DOI: 10.4324/9781315159904-9

plants. In particular, I will apply them to some of the models that were presented in previous chapters of this book. These characterizations allow a more precise consideration of the distinctive characteristics of the various types of models and simulations. In return, they make it possible to better understand their complementarity and connections.

The differences of opinion between philosophers regarding the roles played by models in science are mainly attributable to the diversity of these roles. The disagreements are due, in particular, to the heterogeneity of the examples used in the various arguments. The hypothesis I hope to validate here is the following: this diversity should not necessarily worry philosophy, since order can be found in it by means of a conceptual approach that is gradually refined through induction and broad comparative analyses.[1]

General function, main functions and specific functions of models[2]

It seems necessary, first of all, to distinguish successively between a model's function, nature and principle. Later on in this chapter I will use the term *epistemic function* of an object used in a scientific investigation (a tangible, symbolic or mental object) to denote not simply the function of guarantee, or of aid to accreditation or diffusion that this object appears to perform, from the outside, *for* an already-formed *propositional* knowledge,[3] but also the *form itself of the knowledge* that this thing makes it possible to access. This is because knowledge does not always have a propositional format, even though it tends to rapidly become or be transposed into a propositional format in science. In this context, it is useful to recall that knowledge may take several forms:[4] it may be observational, perceptive or selective/contrastive (experiential knowledge, knowledge by acquaintance, objectual knowledge). Knowledge may also be descriptive, predictive, explanatory, comprehension-knowledge or theoretical (propositional knowledge per se). And, finally, it may also be a form of practical cognition (know-how, procedural knowledge or hands-on knowledge). Only some of these forms of knowledge can be inscribed straight away in a propositional format.

For this reason, the *epistemic function* of an object used in a scientific investigation will serve here to define the *form of knowledge* that it permits us to acquire in the first place. For example, a scientific instrument such as an optical microscope produces a format of knowledge that is initially observational, even though it is possible to cast doubt on the information provided by this observation by recalling that it assumes that the classic laws of optics are valid and that use of the microscope produces no artefacts. It is therefore justifiable to point out that the observation made via the instrument is theory-laden, and that its credibility comes from a series of overlaps between different theoretical representations, different physical principles for the microscopes and different causal interactions with the target system.[5] The knowledge format that the instrument immediately produces, however, remains observational first of all. A model is not an instrument, even though it may be produced by an instrument. It may be objected that certain models are often compared with filters, optical lenses, microscopes or telescopes. In this case, the filter is confounded with the filtrate, and the instrument with the image it produces. The virtual image in a telescope is not the telescope. In the specific case of formal models, the generic schema of the model is confounded with the model itself. A *statistical model schema* whose parameters have not yet been calibrated is indeed an *instrument for reading* an experiment. But once it has been calibrated using parameters with fixed numerical values, it is no longer a schema: it becomes a model – and in this case, a statistical-type model. Like an instrument, however, a model is also used in science because it provides a particular epistemic function. Clearly, of course, a model may even provide several epistemic functions simultaneously on the actual scientific ground, but for the purposes of determining conceptual distinctions, I will initially consider that it provides only one main epistemic function.

What I propose to call a model's *substantial nature* refers to what it is made of: both its substance and its structure. When considered from the point of view of its nature, a model may be a non-living tangible object (for example, a natural or artificial tangible object, an image, or the result of a computer simulation); a living tangible object (for example, a model organism such as a pig or a drosophila, or a model organ that has been removed from its organism but kept alive); a symbolic object (a mathematical equation, a system of equations or iterative rules operating on symbols, for example); or a mental object (for example, the mental depiction of an image or series of images).

Lastly, what I propose to call the *functional principle of a model* is, for its part, the description of how its *substantial* nature (or certain properties of its nature) enables it to carry out its *epistemic function*. A survey of the different forms of modelling suggests that a model's principle may be of four different types: exemplification; denotation; similarity of relationships (analogy); or information compression.[6] To sum up, the model's function is what it does; its nature is what it is made of; its principle is what enables it – in what it is made of – to do what it does.

This first series of distinctions already makes it possible to avoid several types of misunderstandings. Thus, certain taxonomies of models place "theoretical model" and "stochastic model" side by side, on the level of one single alternative. But the qualifier "theoretical" specifies the epistemic *function* of the model, whereas the qualifier "stochastic" specifies a property of the model's *nature* (which in this case is symbolic-mathematical), i.e., a property of what it is made of. The two definitions therefore cannot be situated on the same level of one single alternative. The draft classification proposed by Manfred D. Laubichler and Gerd B. Müller[7] in order to provide a structured sequencing for the collective publication they edited – a publication which, for that matter, is full of very interesting one-off case-studies – exhibits the same type of coherence issues. At the first branch of the classification tree they propose, we find – on one and the same level – a choice between models that at times are "material", and at others "theoretical" or "heuristic". But "material" refers to the nature of the models, whereas "theoretical" and "heuristic" relate to their functions.[8] In other publications there is also confusion, at times, on the level of a single definition. Does "analogical model", for example, refer to the model's *principle* (the fact that it is an analogical relationship that will give it its function), or to its *nature* (in the sense that it is neither numerical, nor even symbolic, but material)? We are often left in doubt. The definition "analogical" for a model is vague and should be avoided unless accompanied by further information.

Let us now examine the general function of models. Until now, the numerous historical, sociological and philosophical works on models have all demonstrated that the nature of a model does not unequivocally determine its function, and nor does a model's principle determine either its nature or its function in an unequivocal manner. On the other hand, they did demonstrate that there was a certain agreement regarding the general function of models. My own classification will therefore start from this minimum basic agreement on the issue of function, but at the same time seeking to clarify it. This will involve proposing a breakdown of the different functions of models based on an analysis that specifies this general function. Given this minimum basic agreement on the prevalence of the functional role of any model, but also bearing in mind the direction of the reciprocal determinations between function, nature and principle, it now seems preferable to start first with the general epistemic function of a model, followed by its specific function, if we hope to then determine more completely its epistemological profile – a profile that should ultimately combine its function, its nature and its principle.

Nowadays it has been fairly widely recognized that the general function of models is mediation.[9] More precisely, their general function is that of *facilitating mediation*, since some mediations create obstacles where none had existed by obscuring certain relationships between the things that they are indirectly comparing. Broadly speaking, a model is expected to provide the function not just of mediation, but of *facilitating mediation*.

What is it, though, that is facilitated by such mediation? I would answer by referring to a characterization of models that is frequently used by computer scientists and some social-science modellers: "To an observer B, an object A* is a model of an object A to the extent that B can use A* to answer questions that interest him about A".[10] In light of this characterization by Marvin Minsky, it may be added that the facilitating mediation of a model is determined, more precisely, according to a double relativity: it is relative both to an observer and to a specific questioning. It should also be noted that the autonomy of the model that is being questioned – an autonomy that has since been recognized in a number of philosophical works[11] – was asserted as soon as Minsky's characterization was made, since the model is described as an object (A*) and not just as a way of saying, a way of seeing, or even as a simple linguistic performance of an analogical or metaphorical nature, as was still the case for Max Black or Mary Hesse, for example.[12]

From the point of view of its general function, it may thus be said in summary that a model carries out the function of *a substitute study object*[13] *in the context of a directed questioning, whose facilitating mediation function is effected between different cognitive capacities or different capacities of apprehension.* For Margaret

Morrison, for example, the main mediation that is sought is effected between the theoretical capacity of apprehension of a knowing subject and that subject's physical apprehension of the real world or, in short, between the subject's theory and his or her data.[14] My classification will include these specific cases, but it will start from the most general function and suggest that scientific practice should give examples of mediation with a more varied range: mediation may take place between many other types of cognitive capacities or capacities of apprehension, and not just for one knowing subject, but also between several subjects. Note that the characterization of a model that I have just set out also makes it possible to grasp that it is a mistake – a category error – to demand that any given model should be simple in an absolute sense. The fact that a model must simplify or facilitate our *access* to another object does *not* mean that the model must be a simple object in itself. It makes no sense, for instance, to say that a model organism such as a pig is *simpler* than a human being. Consequently there is no obligation for a mathematical model or a computational model to always remain small and, as such, simple.

This is not the place to describe in detail the twenty-one specific epistemic functions that a broad comparative analysis of the literature allows us to identify for models.[15] I will present and justify only the five main epistemic functions, then directly provide a table classifying and summarizing the specific functions that subdivide these main functions. Each of the five main functions can be set out based on a breakdown of the general function. They may be identified if we ask ourselves systematically about the different types of facilitating mediation that are not only possible a priori, but that actually exist in scientific practice.

The first main epistemic function resides in the fact that a model facilitates, through mediation, the type of knowledge that takes the form of *sensory apprehension*: namely an experience, an observation or a controlled experiment. It is therefore a mediation between one vague sensory knowledge and another, more informative, sensory knowledge, with the aim of using the latter to improve the former.

The second main epistemic function of models is that of using a mediating object in order to facilitate an initial *intelligible representation* of the target object, i.e., a representation based on concepts and symbols that denote these concepts.[16] This is typically a facilitating mediation between a form of observational knowledge of a target system and a conceptualized form. This mediation first of all requires a standardized formatting (data format, measurement formatting) of some aspects of the experience or experiment. These include, but are not limited to, modellings aimed at the construction of measurements, of data and, in particular, of data models. It should be noted that, unlike the first function, this is a mediation between two different types of cognitive apprehension: observation on the one hand and conceptualization on the other. It should also be noted that conceptualization is not yet the same as theorization. Although it may seem surprising, the models for prediction, explanation and comprehension can already be included in this category of function, in my opinion.

Facilitating theorization falls under the third category of main function. In this context, a *theory* will be considered as a system of knowledge that proceeds by concepts and that makes it possible to coordinate and combine into a single language – whether formal or not – conceptualizations of phenomena that are deemed elementary or fundamental and that belong to a given field, so as to then make it possible to understand or explain through verbal reasoning, deduction or calculation a set, or a whole sector, of phenomena that emerge in this given field of the real. Thus, no model is, strictly speaking, a theory, even though some models are the outlines of theories. By contrast, many different types of models are directly useful for theorization; for such models, this involves, in particular, facilitating the mediation between one or several conceptualized representations (such as data or data models) and the representations of theoretical hypotheses, general principles or theoretical laws. But it may also involve, on the contrary, facilitating the mediation between abstract theoretical hypotheses and figurative representations, in order to make it possible to represent the abstract conceptual relationships of the theory in an observational form.

The fourth main function involves *facilitating the co-construction of scientific knowledge*. It should be noted that although the other main functions concern mediations between different apprehensions or different modes of apprehension of a target system for the same knowing subject, this fourth main function consists of facilitating the mediation between different apprehensions that are specific to different knowing subjects, or to different scientific communities, or else, more generally, to different scientific disciplines. Fairly logically, this inter-subjective or interdisciplinary mediation function has been especially emphasized by socio-logical and socio-historical approaches to models, whereas the first three functions have been studied more specifically by philosophers of science.

The fifth and final main epistemic function of models includes the category of mediations between either conceptualized or theorized forms of knowledge, on the one hand, and practical forms of cognition (actions, procedures, know-how) on the other. This function includes all the models that are aimed primarily at *facilitating decision-making and action* (whether individual or collective) rather than at representation. This main function has also been studied and emphasized by sociologists of models above all.

General characterization and classification of computer simulations

Like the term "model", the term "simulation" is polysemous. Nevertheless, an order can be identified within this polysemy. It should be noted, first of all, that the Latin-origin suffix "-ation" indicates that the term is not primarily used to denote an object, such as a model, but instead a process: "model" and "simulation" are therefore not on the same semantic level. A large number of words that are constructed along the same lines also denote the result of the process, however. The word "bifurcation", for example, may at times mean the process and at other times the results of the process. For this reason, it is sometimes appropriate to refer to a model object by means of the term "simulation". In order to avoid confusion between the process and its outcome, which is very common in the literature and – to my mind – reduces

Synoptic table of the functions of models[17]

Main function	Specific functions	Examples[18]
I Facilitate sensory apprehension	1 Make certain properties visible on a substitute	Specimens, samples, wax écorché models
	2 Make certain relationships visible on a substitute	Diagrams, maps, model organisms, model boats
	3 Facilitate memorization through ordered representation	Rhymes, images, memory theatres, local memory[19]
	4 Condense information in order to facilitate sensory recollection at will	Symmetry axis systems, analytical use of statistics (average, variance, etc.)
II Facilitate intelligible formulation	5 Facilitate data compression in order to prepare for conceptualization	Data models, statistical models for synoptic use, distribution curves, statistical envelope
	6 Facilitate the choice of types of entities or properties	Conceptual models, knowledge model, classification, hierarchical or ontological models
	7 Facilitate reproduction of data structures through the use of intelligible means of deduction or calculation	Digital models (digitization of scenes), phenomenological models (data-based), descriptive and/or predictive models, design models (engineering models), models reconstructing data via spectral data analysis
	8 Facilitate explanation[20]	Simple interaction model, finite interaction sequence model, mechanism model
	9 Facilitate comprehension[21]	Equilibrium model, optimality model, aggregate-level variational principle model, thermodynamic bifurcation model (far from equilibrium), attractor, topological model (e.g., morphogenetic landscape)
III Facilitate theorization	10 Facilitate creation of a draft theory: theoretical model	Set of axioms and hypothetical fundamental relationships conceived as a draft theory (whether a theory for explanation or a theory for comprehension), model of hypothetical axiomatization of developmental biology
	11 Facilitate the interpretation of a theory: model of the theory	Mental images representing theoretical entities, physical models of mathematical theories

(continued)

(continued)

Main function	Specific functions	Examples[18]
	12 Facilitate the illustration of a theory: model for the theory	Model of currents in fluid for electrical current flow, model of electrical oscillators for a mathematical theory of population dynamics
	13 Facilitate an internal coherence test of the theory	Semantic model, concrete model (i.e., which refers to objects) of a completely formal theory, model of truth values in propositional logic theory, Euclidean model for geometry
	14 Facilitate the applicability of the theory	Approximate semantic model, empirical sub-structures[22]
	15 Facilitate the calculability of a formal theory	Partially phenomenological or approximate model of the functioning of a mathematical theory, calculable discretized model of a mathematical model, *computational template*,[23] Type 2 simulation model[24]
IV Facilitate the co-construction of knowledge	16 Facilitate the hybridization and co-calculability of several theories	Mixed model in physics and chemistry (e.g., polyphase model), ad hoc model, asymptotic model, multi-scale model
	17 Facilitate communication between scientific actors	Database, open ontology, popularization model
	18 Facilitate consultation and collaboration between stakeholders	Multi-aspect model for consultation, common scenario-exploration models
	19 Facilitate the co-construction of representations and means of verification of mixed systems (human/non-human)	Action research model, participatory model, companion modelling
V Facilitate decision-making and action	20 Facilitate action *on* a mixed and heterogeneous system	Decision model, decision trees, crisis-management model, decision-making heuristics
	21 Facilitate decision-making on action *in* a principally notional system	Market forecast model, financial derivatives modelling

the intelligibility of the argumentation, I propose to use the terms *p-simulation* to denote the process and *r-simulation* to refer to its results.[25]

I will now suggest a minimal characterization of the nature of a p-simulation. This characterization will be adapted to computer simulations. Its validity for other types of simulations can also be demonstrated, although I will not do so here. It appears to me that a p-simulation can be considered to be above all a symbolic process, or more precisely a *process of transformation of symbols*.[26] An r-simulation, therefore, is the result of such symbolic transformation. Since a p-simulation is not primarily characterized by its epistemic *function*, but rather by its processual *nature*, it is easier to understand why an r-simulation at times may have, but doesn't necessarily need, an epistemic function of the same type as a model's.

It therefore cannot generally be said that every simulation is a type of model, nor that a simulation is always the processing of a model. It is also not always true that a simulation is a model immersed in time, as frequently seen in the literature, if by "simulation" we mean the result of the process. It is possible, for example, to make a p-simulation construct an r-simulation of a mathematical model of probability distribution by means of a Monte-Carlo technique. Although, of course, the p-simulation takes time (the time for computation), the resulting r-simulation nonetheless has no specific temporal dimension: it may be stocked, transferred and used *en bloc*. A simulation therefore does not always intrinsically have a temporal dimension. Thus, even though this is often the case, a simulation is not always "a time-ordered sequence of states that serves as a representation of some other time-ordered sequences of states".[27] Here, too, it is important not to confuse process and result. The relationships between models and simulations are thus rich and complex.

To further investigate this point, it is necessary to shed light on the nature of the procedure of symbolic processing that is carried out by a p-simulation. In order to avoid unduly complicating the text, I will use the term "simulation" when referring to a "p-simulation" from now on – unless otherwise specified. If the current diversity of types of computer simulation (numerical, using Monte-Carlo techniques, rule-based or object-based) is taken into account, it may give the idea that the symbolic process implemented by a simulation always takes place in at least two distinct steps.

During the first so-called *operational* step, operations on symbolizing things or on symbols of things take place. These operations take the form of regulated interactions between these things or symbols of things. These things or symbols of things are presumed to always refer – either by exemplifying or by denoting – to elements, properties or real or fictive aspects of a target system that is, itself, either real or fictive. It should be noted that the symbolic function of a thing or a symbol does not disappear if what it refers to does not actually exist. Like the noun "unicorn" in day-to-day language, the name of a finite element involved in a numerical simulation is a denoting term that has no denotation, no referent. This is the fundamental reason why neither this step of the simulation, nor the whole of the simulation process, can be defined as a particular case of modelling: for a simulation to occur, there may be no actual target system (as is the case when object A in Minsky's characterization does not strictly exist, in the sense that it is not even considered to be fictive), but there will nonetheless still always be a regulated interaction between symbols. This interaction is called "computation" in the case of computer simulations. But computation, in this context, has a broad meaning and does not necessarily denote a calculation.

The second step is called *observational*. The rendering, visualization, measurement, or any other reuse of the global results (or patterns) of the first step is carried out during this step. These results themselves are considered as new symbols that denote or exemplify the potential target system, if there is one. They refer in an external manner – again, either fictively or in actual fact – to the real or fictive target system, albeit on a different denotational level than the one referred to by the elementary symbols of the first step. Furthermore, it appears that these resulting symbols also refer in an internal manner: i.e., they also refer to the elementary symbols that interacted in the first step. Just as the name "x" given to the location variable of a moving object in mechanics may be used to denote the elementary symbol "dx" of an elementary movement of that moving object, so the overall symbols resulting from the interaction and aggregation of these elementary symbols generally denote the elementary symbols that they are composed of, and these elementary symbols, in return, exemplify them. A simulation thus uses the descent down certain denotational hierarchies[28] in order to replace the combinations that ought to operate at a higher – but unachievable – symbolic level with exemplifications of symbols and operations that take place at lower symbolic levels, followed by rises up the denotational hierarchies, and then by direct measurement procedures on the resulting overall symbols.

It should be noted that the exemplifying symbols that I am referring to here obviously have no material substance. This is a *relative exemplification*.[29] Using Goodman's terminology, it may be said that they exemplify because they are at a lower level in the denotational hierarchy in question here than the term that denotes them. The debate around the essential or non-essential nature of materiality in the experimental dimension of knowledge, and in particular of simulations,[30] can perhaps be broadened if we concentrate not just on the general aspects of models, experiments and measurements, but also on these low-level, diverse and evolutive relationships between the elementary symbols that are at the heart of computer simulations.

Nevertheless, Margaret Morrison is right to point out that the process of measurement – which corresponds in this instance to the final sub-step of step 2 – is important for assessing the almost experimental nature of a simulation. A simulation, as in the case she cites of a simulation of particle collision,[31] makes it possible to replace a mathematical procedure known as a *problem inversion* when there is no procedure that is analytically calculable. In Chapter 6 of this book I analysed this function of data inversion by simulation, allowing the construction of a credible *explanatory model* by analysing the case of forest supra-simulations. It should be recalled that such supra-simulations have been carried out since the 1990s in order to interpret satellite maps of wooded regions. I would add, however, that the dual fact that, on the one hand, an r-simulation makes it possible to *give us access* – by means of its mediation – to a measurement that we would otherwise not be able to carry out directly on an inaccessible target system, and that, on the other, the manner in which the r-simulation indirectly gives us access to this target system *also* involves *a measurement process*, does not mean that the first measurement process is literally replaced by the second. Nor does Margaret Morrison make such a claim, for that matter. But it is true that a form of iconic resemblance between the two practices is seen here as being impressive and appearing to make the mediation even more persuasive. This iconicity is nonetheless the result of a long sequence of mediations. All that can be said is that one process of measurement is mediated by another process of measurement.

In this regard, there is a fundamental special case that is often forgotten in the epistemological literature; this is the case of simulation models of complicated artificial mechanical systems, such as an airplane. Von Neumann set out an extremely convincing argument on this subject, which I have cited several times.[32] In essence, he demonstrates that it will inevitably become preferable to carry out measurements on a discretized simulation of this type of complicated system because, sooner or later, unlike a target system that is measured directly in ways that will necessarily remain partially analogical (i.e., not fully discretized), the *quality of the joint control of the representations, of their interactions, of the measurements of the results of these interactions, and of the associated calculations* will exceed the quality of similar controls carried out on the actual object (the prototype). This explains why, in the case of complicated manufactured systems for which the dimensions are very precisely known and controlled in advance, as in the case of aircraft, simulations have for many years been much more reliable than experiments in terms of measurement: the simulated system is much more reliable from the point of view of the empirical knowledge it gives us than the actual system itself. It should be pointed out that Von Neumann's argument is based on a rather detailed consideration regarding the *joint control* of data models and measurement models, and not on consideration of just the methods of representation and measurement. Of course, not all target systems are of this type; consider, for example, physical systems that are not completely manufactured, or living or social systems. This is why, insofar as these systems are concerned, the issue of the reliability and empirical value of measurements on discretized simulations still arises.

It is important to remember one thing, however, regarding any simulation: in order for a process of regulated interactions between symbols to be a p-simulation, it must be constantly accompanied by a rerouting of the reference for its resulting symbols, not only with regard to their external references, if there are any, but also to their internal references. In other words, there must be changes in the levels of reference (denotation or exemplification) of these resulting symbols, in view of the denotational hierarchies that our background knowledge establishes between things and symbols, in the context of the scientific investigation in question.[33] By studying in greater detail and then specifying this essential function of both internal and external reference, we can distinguish the three principal types of computer simulation.

The first type of computer simulation is what can be called a *model-led simulation* or *numerical simulation*. In this case, we start with an existing mathematical model, which is often in the form of an equation and is analytically intractable. The variables of this model are discretized in order to produce discrete elementary symbols (finite elements or finite differences). This discretization process determines the direction of the route of reference leading to these elementary symbols: it goes towards the bottom of the model's internal denotational hierarchy. The first step of p-simulation then organizes the interaction between these elements, and the second step organizes a rise back up the internal denotational hierarchy, together with a measurement of the resulting symbols. The result of this measurement process is often considered to be the approximate result of a calculation, and not the result of an approximative calculation, since numerical simulations are generally based on convergence theorems. A numerical simulation therefore replaces an analytically intractable deductive calculation by a descent in the denotational hierarchy and a series of step-by-step interactions, followed by a rise back up the hierarchy and a measurement. It should be noted that a numerical simulation therefore has no external target object in the sense of Minsky's "object A". This is why, even though it is indeed a process of symbolization, a numerical simulation cannot be considered as a type of model. As a process of (internal) symbolization, however, it can justifiably be said that we *apply* a numerical simulation to a model and therefore that *we are simulating a model.*

The second type of computer simulation is *rule-based* simulation or *algorithmic simulation*. Unlike numerical simulation, algorithmic simulation is not preceded by a preliminary descent in the denotational hierarchy. Instead, it begins directly with the imposition of external reference relationships between certain elementary symbols and certain elementary rules of interaction between these symbols on the one hand, and certain properties of the actual or fictive target system that is the subject of investigation on the other hand. Nevertheless, the fact remains that algorithmic simulation also then proceeds in accordance with the same two steps: an operational step that organizes the regulated interactions between symbols, and then an observational step of rendering, visualization or measurement of the resulting patterns, with a change of reference levels. The observational phase may have some surprises in store, since often for this type of simulation there is no theorem demonstrating beforehand that there will be convergence or robustness of the results.[34]

The third type of computer simulation is *object-based* or *software-based simulation*. This is used when there is no single theory or axiomatically unified mathematical model of the target system, nor even a system of rules that are sufficiently homogeneous or that deal with the same type of aspect or on the same scale. A good method is then to "objectify" certain elements of the target system. These elements are chosen not because they have been deemed fundamental, as would appear to be implied, however, by a theory-based approach, but because the modellers consider that they are key points around which the computer can make the computations of differently formalized and axiomatized sub-models compatible, one step at a time. These target-system elements are then linked with computer objects, in the sense that the word "object" possesses in the so-called *object-oriented* programming languages. The benefit of objectification is that it makes it possible to prolong and amplify the process of de-abstraction that underlies rule-based simulations, but without being limited to just one type of de-abstraction at a time. The different attributes and the different methods (rules of interaction between objects) are specific to different symbolic objects that are numerically as well as qualitatively distinct. Multi-Agent Systems can be included in this type of simulation.

System simulation, model simulation, system-simulation model and model-simulation model

The foregoing analyses allow us to clarify certain relationships between models and simulations. It becomes possible, in effect, to explain the differences between system simulation, model simulation, system-simulation model and model-simulation model.

A *target-system simulation* is actually an r-simulation, because it is intended to designate the result of a symbolization process that is, for that matter, of p-simulation type. This process does not necessarily take place through the use of a computer, but the symbolic treatment does, indeed, exist here in the sense that the r-simulation refers to (i.e., denotes or exemplifies) all or part of the target system: the operative and interactive phase exists,[35] but it is the observational phase that is most important. The manner of referring often demands a partial iconicity, i.e., a similarity of nature. Thus, the temporal iconicity is often considered essential in an r-simulation of a system,[36] but this essential nature may be entirely incidental, or it may be absent, or even feigned. Using my chosen rhythm of internal clock, I can make a computer temporally reproduce the result of an earlier p-simulation that I had stored beforehand in its entirety in the memory in the form of an organized but atemporal table. The effect of simulation, in the sense of a fiction, would be doubled up in this instance, because the time of this pseudo r-simulation would not be the same as the time of a p-simulation. An r-simulation can therefore be deployed in time at will without being the actual deployment of a p-simulation process. Thus, in Chapters 3 and 5, I demonstrated that the AMAPsim software, for example, could produce an r-simulation of the dynamics of a system (an r-simulation of the "growing tree" system) by means of *ex post* reconstruction and using a stop-motion technique, without constantly relying on a p-simulation by dynamics, although this was what the AMAPpara software did. As we can see, what we are looking for in the temporal dimension of an r-simulation is that it should be able to refer iconically to the temporal dimension of the target system as seen from a dynamic perspective, for example. For this to occur, however, it is not always necessary that the p-simulation process that it derives from should itself be equally iconic in terms of the target-system dynamics. Furthermore, it should be recalled that an r-simulation, by nature, does not have to possess a temporal dimension.

In the scientific literature, the expression *model simulation* refers to the computation process that affects a formal model. The process takes time (the computation time) and it is valid to say that a model simulation is the fact of immersing a formal model in time. This model may be of a logical, mathematical or software-based nature. An internal symbolic processing is carried out on the model, and the result of the aggregation of elementary computations is examined. This is typically the case for discretization and for all the various approximation strategies and strategies for descent in the denotational hierarchy that are used in a numerical simulation.

A *type 1 simulation model* (simulation model-1) refers to a formal mathematical or computational model that is conceived in order to produce an r-simulation of a target system. It is a *system-simulation model (SSM)*. In this case, the term "model" is used to designate the "mould" that unites the equations or other formal relationships between symbols that enable a computation process to take place in order to ultimately produce an r-simulation of the system concerned. This simulation model is indeed a model; it is a model *for* an r-simulation. In effect, it plays the role of mediation between the formatted data, the hypothetical mechanisms and the conceptualized theoretical hypotheses on the one hand, and an r-simulation of the target system on the other. It is therefore a mediator for another mediator, and it is by participating in this sequence of mediations[37] that it is able to indirectly carry out epistemic functions Nos 7, 8, 10 or 11 for the target system itself, whereas the r-simulation that it produces may be seen as a model that, in this case, directly provides functions Nos 1 or 2 for the target system.[38]

A *type 2 simulation model* (simulation model-2) designates a transformed state of an earlier formal model. In this instance, the formal model that is to be transformed is an approximation of a theory or an outline theory, i.e., a formal model with function No. 10. The formalized theory or the theoretical-mathematical model therefore remains a necessary starting point. This is often possible in physics or chemistry, where many theories are available. The *type 2 simulation model* is then designed so that the p-simulation of the *theoretical-mathematical model* can be carried out by computer. It is thus a *model-simulation model (MSM)*, and should not be confused with a *system-simulation model (SSM)*. It is primarily this type of simulation model, together with its variations, that is precisely described in the work of Eric Winsberg.[39] Fairly often, the simulation model takes the form of a computable algorithm or computational model, which Winsberg also calls a "solver",[40] or Paul Humphreys calls a "computational template".[41] Winsberg, however, also demonstrates that a wide range of modifications of the formal writing of the initial model can be found. According to him, there is an internal hierarchy between different types of more or less approximate simulation models for any given theoretical-mathematical model.

It seems to me, however, that in the hierarchy Winsberg proposes (mechanical model centred on objects, dynamic model including the values of particular parameters, ad hoc models that also include parameterizations, discretized model and, lastly, approximated visualization model of the phenomenon),[42] there are confusions between model simulation, model-simulation model (discretization), system-simulation model (object-based model) and r-simulation (visualization). Indeed, as I tried to demonstrate in this book for the case of plant-simulation models, as far as complex simulation models and the differentiated validation procedures that accompany them are concerned, a complex hierarchy, or more precisely a tree-diagram, may admittedly appear within the simulation models. But the model trunk – if we adopt the analogy of a tree-diagram of models – is then a type-1 simulation model or SSM. This cannot be a theoretical-mathematical model or even a *simulation model* of such a model (MSM). In fact, as soon as a simulation model resorts to even one single parameterization, i.e., to the forcing and inclusion of an empirical ad hoc relationship that is valid at a different denotational level than the levels for which the initial theoretical model was validated, as soon as it anchors its parameters to a meso-empirical level by anchoring them directly to different data than just the data that sustained the initial formal theory, or even as soon as a computation model relies on a formal collage on a meso-scale between theoretical models of a handshaking module[43] type, then it is no longer a *simulation model of some initial theoretical model (MSM)*, but instead a *complex model for the target-system simulation (SSM)*.

The SSM category is not simply a mirror-image of the MSM category, however. To some extent, the SSM is more epistemically potent, because it includes the possibility of integrating together model simulations and MSMs, as well as SSMs. The issue of validation of such models is therefore complex, but it is necessary that the type of distinction I am highlighting here should be made, and that thought be given to the consequences of these distinctions with regard to the problem of validation.[44] Thus, it is a conceptual confusion between the different types of simulation models that leads to making an issue of this idea that the fragments of completely fictive models – or "false" models, given the reality – or even fragments of models that are directly "contradictory" to the initial theory, seemingly have this mysterious ability, despite everything, to give correct and useful results.[45] Since system-simulation models (SSM) allow us to make the computer manipulate something other than propositional-format knowledge (for the requirements of step-by-step co-computation), the question of the "truth" regarding these SSMs remains valid in one sense, of course, but it must be expressed differently and much more precisely right from the start, i.e., by taking account of the epistemic heterogeneity of the forms of knowledge that are simultaneously dealt with by these simulation models. This question of the truth cannot be posed head-on, or in a unique and uniform manner. The multiplicity of ways that symbols refer to – or are anchored to – target systems and to the background knowledge we have about them must be recognized first if we want to better understand the complexity and richness of the validation procedures of these current complex simulation models.

On this point, the results of my work on distinguishing between types of simulation models are in agreement with, for example, the proposal advanced by Sergio Sismondo. Sismondo asks that the importance of the realism of knowledge arising from approaches such as that of natural history should be recognized:

> Natural historical knowledge can thus be considered empirically realistic in the sense that it describes nature as it is found, as any skilled observer would find it. No twists or turns of interpretation are required to say that natural history knowledge is about nature, even while it contains abstractions, imposed classifications, simplifications, and so on.[46]

An approach of this type – but which must be defined further still – could in part complete the propositional approach inspired, in its turn, by the use of models and simulations in physics. The epistemic role of images should also be more widely recognized and is, for that matter, the subject of more and more in-depth research.[47]

In point of fact, the contemporary use of complex simulation models demonstrates the compatibility between classic formal models with functions Nos 8, 9 or 10 with an observational and descriptive science approach, and thus with models with epistemic functions Nos 1 and 7. The result of this mixed nature in this type of complex simulation models is that it imposes specific forms of validation that are both multiform and multi-scale, and which, for that reason, are different from what a uniform validation regarding the truth of solely propositional knowledge might be.[48]

Applications to different plant models and plant simulations

Having presented these conceptual distinctions and classifications in a somewhat abstract manner, I will now illustrate them by means of examples. I will do so by redefining some of the different types of models and simulations presented in my longitudinal and comparative case study.[49]

If we return first to one of the main sources of the formal model method in biology, we find the name of Ronald A. Fisher. Compared with the physicists of the preceding century – for whom a model in the mathematized sciences was still only a real or fictive physical model that served either to interpret a mathematical theory (Boltzmann, model function No. 11) or to illustrate it (Maxwell, function No. 12) – Fisher introduced a new type of model:[50] formal-type models. The epistemic function of these models was also different: their purpose was to analyse the experiment and to condense the information drawn from it (function No. 4). They could also be used to reconstruct these data (function No. 5), as was subsequently highlighted by Jerzy Neyman.

Next, the laws of allometry (Huxley, Teissier), called allometric growth models from the 1950s despite their mathematical nature and their equation format, appear clearly as phenomenological models. They are descriptive and partially predictive, but they are not explanatory, and – although this has been highly debated – they are also not based on a consensual theory that enables understanding of the underlying processes. As a result, allometry models provide no more than function No. 7[51] in the study of plant growth.

In turn, the general axiomatization of growth and of the morphogenesis of living beings, as proposed in Joseph H. Woodger's formal work, could be considered as an attempt at theorization.[52] As a weak attempt, which met with little support for that matter, it may be considered to go no further than a theoretical model (function No. 10). It should be noted that, although it is not a theory, this formal model is nonetheless neither an interpretation (function No. 11) nor an illustration (function No. 12) of a theory. It therefore clearly represents a distinct category of function.

In 1952, when Alan Turing published his chemical-mathematical model of morphogenesis in order to explain the emergence of dissymmetrical patterns in a homogeneous substrate, he introduced a number of different things. First of all, he introduced a mathematical-type model, in an equation format. But these non-linear equations were introduced while at the same time focusing on the fact that their construction was based on the credibility of a type of interaction mechanism that could, for that matter, be easily visualized by the unaided human mind: the mechanism of reaction-diffusion. Thus this model aimed to provide a primarily explanatory function (function No. 8). Since the approximate calculation of this model also made it possible to roughly reveal the qualitative appearance of patterns observed in the phenomena of morphogenesis in living beings (spots on the fur of certain animals), this mathematical model may be considered secondarily to provide function No. 9 (comprehension). Furthermore, this model is not the outline of a complete or unified theory, since it is based only on certain laws of certain chemical and physical theories. It is therefore not a theoretical model (No. 10) as such, nor a model for the interpretation (No. 11) or illustration (No. 12) of a theory. Turing's model is also known for having been one of the first to be processed by computer, and is often referred to as a simulation. Turing did, of course, carry out a *numerical simulation* of this non-linear mathematical model: he did so because it was necessary in order to solve the model. The 1952 article, however, did not publish a *simulation model* strictly speaking, nor a *target-system simulation*.[53]

It was a completely different story for Stanislaw Ulam's first simulations of plant growth. Turing had discretized the continuous variables that represented the substrate only at a later point, in order to make it possible to simulate his model numerically.[54] Ulam, however, represented the substrate directly in the form of a lattice, i.e., in a form that was both discretized and spatialized, in two dimensions. He then used the computer to apply what would later be known as the technique of cellular automata: the iterated interactions between the cells of this lattice. In this respect, Ulam advanced one of the first forms of *system-simulation model* for plant growth. It was considered, however, that the triangular cells that he used did not iconically represent living cells, but rather pieces of individual branches. The individuals denoted by this simulation model were less abstract and fictional than the finite elements in the numerical simulation of Turing's model, but they remained partially fictive since they were poorly identified and characterized from a biological point of view. At the end of its computation this simulation model produced an r-simulation that nonetheless closely resembled, broadly speaking, a plant-type branching shape. This simulation model therefore indirectly provided a No. 2-type epistemic function: making certain relationships roughly perceptible. By so doing, it indirectly made the overall form of certain potential mechanisms of interaction between the growth processes of different organs more credible. Thus it also provided function No. 8 (explanation), but did not enable any prediction (No. 7) or comprehension (No. 9), and was not theoretical (No. 10).

Around the same time, Murray Eden used a plotter to represent a process of biological growth, which this time was valid on a cellular scale.[55] A living cell was represented by a point and its geometric coordinates were shown on a plane. At each step of time in the p-simulation, by conforming to a mathematical law of probability that in turn was simulated by a Monte-Carlo technique, each cell could give rise to a daughter cell in its immediate vicinity within the lattice. The idea was to watch this patch of growth increase on an aggregate scale, i.e., on a different scale than the cells, and to evaluate its realism in terms of empirical observations. This was one of the first forms of direct *algorithmic simulation* of multicellular development or of bacterial cultures (type-2 simulation). By basing itself on the function of mediation and credibilization that is achieved through visualization of the patch of growth (function No. 2), the *system-simulation model* at the heart of this *algorithmic simulation* indirectly provided a function of possible explanation (No. 8) of certain forms of growth or of cellular multiplication. It should be noted that it was thus Eden, rather than Ulam, who explicitly introduced here the stochastic element (by means of Monte-Carlo techniques) in computer simulations of biological growth.

Subsequently, as we saw in Chapter 1, what was still just a simulation of a spatialized combinatorics in Eden, was made more biologically realistic on the scale of branching plants by Dan Cohen. In order to reveal the dissymmetries, preferred growth directions and branching shapes, Cohen bent the laws of the probability of birth of cells by polarizing the geometric space through the simulation of a "morphogenetic field" in the discretized geometric plane. He ended up with ramified threadlike shapes directed upwards. The points no longer denote the cells, however, but rather entire organs such as branches. Eden's *model for an algorithmic system simulation* thus became a model for an *object-based simulation*. The objects were more abstract from a biological point of view, but they were chosen for a specific result: it was on the scale of these objects that the iterated stochastic rules of the computer cells, on the one hand, and the geometric rules of the morphogenetic field, on the other, could *interact*. These two categories of rules were no longer co-calculable in one single theory, or even in a simple interpretative (semantic) model of a single theory. The r-simulation of this simulation model could, of course, have become biologically realistic on the scale of branching plants. But, because of the change of scale of the objects that are denoted in the target system, the mechanism of growth or ramification for each piece of branch was represented phenomenologically (empirically) and in a much less explanatory manner than was the case for the biological cells and their mitosis mechanism. Consequently, even though Cohen's simulation model produced an r-simulation that was more and more visually realistic (function No. 2) and descriptive (function No. 7), the price to pay for this improvement was clearly that it provided less and less of function No. 8 (explanatory). The same was true of the *algorithmic p-simulation models* for the *geometric r-simulation* of trees (calibrated on a large amount of real data) such as those that Honda and Fisher later produced: they would further improve Cohen's model, but by following the same principle of increasing only the geometrical realism. This type of *system-simulation model* can therefore be considered as ultimately having only one single indirect epistemic function, function No. 7 (phenomenological): this is a forerunner of some contemporary computational models that are a matter of pure data mining together with pure data reconstruction.

In Chapter 2 of this book I showed that Lindenmayer had deployed his own logical model, inspired by Woodger, precisely in order to avoid this drift towards purely phenomenological simulation. His aim was to implement a model that would be both explanatory (function No. 8) and predictive (function No. 7). This is why I described it as a *logical model*. Compared with the geometric branching approaches of Cohen and Honda and Fisher, its main innovation is conceptual in nature: it is based on the representation of cellular mitosis in terms of a formal rewriting grammar. For the first time, however, this type of grammar could now, with precision and bit by bit, deal with not just mitosis and its mechanics but also, from a logical point of view, the sequence of cell types in the cells' lineage. In this way, it also allowed a complementary handling of ramification in a represented geometric space. The first *algorithmic simulation models* produced by computer that Lindenmayer proposed for algae are therefore once again both explanatory (No. 8) and predictive (No. 7) at the same time, albeit only for certain categories of algae. The criticism that Brian Carey Goodwin directed at these models, however, was that they were not equally theoretical (No. 10) in the sense of an authentic biological theory based on the representation of more fundamental biological processes and consequently of broader applications. According to Goodwin, the algorithmic simulation model was purely ad hoc and, for that reason, was phenom- enological (solely function No. 7), whereas, for Lindenmayer, this model could even be considered theoretical (No. 10) in the sense that it not only grasped in an explanatory manner (No. 8) but also (since his theory was simply a set of axioms of the rules of rewriting) completed the logic of the cells' reproductive behaviours, which of course took place at a level above the biochemical interactions, but was nonetheless real, and truly biological, for all that.

Chapters 3 and 4 of this publication then demonstrate how, some time after the aforementioned works, a researcher whose main and fairly constant aim was to make the biometric models of higher-plant fructification more effectively predictive (func- tion No. 7) was induced to explore the other epistemic functions that his models also needed to provide in order to achieve the desired predictive function. Thus, in these chapters, we first saw Philippe de Reffye initially resorting to the solely analytical uses (function No. 4) of the multivariate analysis models resulting from Fisher's work. This turned out to be in vain, however: like others before him, de Reffye discovered the highly non-linear nature of the fructification processes. As a result, neither the linear statistical analysis models (function No. 4) nor the allometric-type phenomenological models (function No. 7) worked. One solution was to try to repro- duce the tree's growth and fruit-bearing in a more botanically faithful way, and at the same time to make a more synthetic, rather than analytical, use of the probabilities (function No. 5). It was the need to evolve towards this synthetic use of probability laws (a use based on Monte-Carlo techniques) in the growth, branching and fructifi- cation simulation models that led de Reffye to truly move from *mathematical models to computer simulation*.

The simulation models (SSM) that de Reffye progressively constructed required that the step-by-step computation should follow the tree growth more closely from the point of view of their botanical realism. De Reffye then demon- strated that, in the case of higher plant growth, in order for the final r-simulation to be predictive, the p-simulation had to be the most realistic possible at each step, at least at the level of the operation of growth and branching of those individual key points: namely, the buds. If the simulation model was to be recognized as being predictive, then the series of intermediary r-simulations obtained by computation of the model would have to constantly provide the function of representation (No. 1) and the function of making the measurable relationships perceptible (No. 2) by means of a simulated 3D image.

In order to do this, however, it was also necessary for the simulation model to take account of and explain (function No. 8) the simultaneous and integrated processing of several mechanisms with different characteristics: physiological, geometrical, topological, etc. It had become clear that, in order for these tree-growth and fructification models to correctly carry out function No. 7, it would be necessary to improve function No. 8 of the simulation model, and the hybridization function (No. 16) of the theoretical models, as well as providing functions Nos. 1 and 2 on a step-by-step basis. When de Reffye made his simulation model more universal, it was indeed at the scale of the buds that his simulation models (which were by then object-oriented) simultaneously processed several formal constraints that had emerged from differently axiomatized sub-models. This was the age of fragmented and pluralistic modelling that emerged from what I called the *pluriformalized and multi-process approach*. The validation of the simulation model then took place not only on the basis of a comparison between field data and the final r-simulation, but also on the basis of a series of longitudinal and multi-scale comparisons between the field dynamics and the representable steps of the p-simulation. The validations were thus pluralistic in terms of knowledge formats, multiscale, "multiphysics", multi-process, and for this reason were cross-validations. The step-by-step intertwining of sub-models and the composition of their validations nonetheless remained fairly transparent and could be broken down again if necessary. In this way, unlike the cases of complex climate models studied by Winsberg and Lenhard[56] in which the integration was so great that it became inextricable, there was less risk of confirmation holism. Furthermore, the dependence on the specific technical route that had been followed (entrenchment) was greatly reduced, although it undeniably still existed.

Chapters 5 and 6 relate how this type of computer-simulation model later revealed itself to be a particularly open and evolutive modelling process. It fostered the successive integration of forms of knowledge arising from different disciplines, such as computer graphics to begin with, and later forestry, with the integration of empirical laws such as the law of water-use efficiency. The case of the development of supra-simulations then demonstrates the capacity for combination and accretion that may be possessed by r-simulations that are produced by such models. Thus a bud-by-bud r-simulation of an individual tree may in turn be used on a higher scale as a dynamic elementary symbol of a new p-simulation in order to then produce an overall r-simulation of the forest. This use further enriches the sequences of mediation between that which enables the simulation model to provide the No. 8 function of explanation (the mechanisms affecting the buds) and that which enables it to indirectly provide the No. 2 function of presentation in an observational form (as a 3D image of the simulated forest). In principle, it also further complicates the problem of the justification and reliability of the simulations. But it is precisely the computerization, as well as the calibration on field data of complex digital scenes that are also partially explanatory in origin, that allow an unprecedented control of the reliability of these long sequences of mediation.

In the face of this growing complexification of simulation models, I demonstrated in Chapter 7 that *remathematization* may become desirable, in particular in order to reduce the number of computation steps, but also to facilitate the problem-inversion techniques that are necessary for making the search for optimal agronomic applications more reliable and systematic. This is a process that is becoming possible and that is also an indication of the maturity of a particular domain. Remathematization involves reducing the number of mediations in the sequence of mediators, i.e., reducing the number of simulation phases per se (interactions between elements, followed by observation). This is only possible once the simulation model has been correctly calibrated to be both explanatory and predictive at the same time. An intermediary mathematical model is then sought by induction on the r-simulation or else on certain specific phases of the p-simulation itself. A partial simulation model is thus carefully replaced by a more abstract partial mathematical model. The latter will take its place in the overall SSM. The epistemic function of the partial mathematical model found on this secondary ground may be either purely phenomenological (function No. 7) or explanatory (function No. 8). As far as the remathematizations at the root of the GreenLab model are concerned, they were first found empirically on simulation models, but the biologist-modellers found, a posteriori, that they were also biologically explanatory since they represented the behaviours of groups of organs that had, until then, been neglected by direct symbolization – namely the metamers – but which in fact turned out to be homogeneous and significant from a biological point of view. In the face of this extraordinary case, it is clear that we cannot accuse every detailed simulation of making us relinquish comprehension and definitively surrender ourselves to big data and the pure description of reality, without any further search for conceptual comprehension on top of that. On the contrary, some detailed integrative simulations, such as those I have described here, become an unheard-of field of exploration in the search for new models of comprehension. Such simulations are even, no doubt, the only field of exploration that is still capable of making this possible for certain complex objects. A conceptual epistemological analysis paired with a comparative history of science makes it possible to highlight this diversity in the contributions of simulations and especially the innovative nature of some of their contributions. Thus we can see that a detailed integrative simulation may at times lead us to perceive and process data and sub-models on other scales and from other perspectives. It is this that, in return, may foster the conception of further models that are aimed as much at explanation as they are at comprehension, and that are not aimed solely at description or prediction.

Notes

1 My inductive and comparative philosophy of science approach is akin, for example, to what Renate Mayntz calls for in the specific case of simulations: "It is likely that the relative importance of simulation differs among disciplines, but to be able to state such differences would require a comparative empirical study", in: "Research technology, the computer, and scientific progress", G. Gramelsberger (Ed.), *From Science to Computational Sciences*, Zürich, Diaphanes, 2011, pp. 195–207; p. 200.

2 The classification proposed here adopts and summarizes the attempts at classification that have already been published in Varenne (F.), "Fragmenter les modèles: simulation numérique et simulation informatique" [Fragmenting models: digital simulation and software-based simulation], in P.A. Miquel (Ed.), *Biologie du XXIème siècle – Évolution des concepts fondateurs* [21st century biology – evolution of the founding concepts], Brussels, De Boeck, 2008, pp. 265–295; Phan (D.), Varenne (F.), "Agent-based models and simulations in economics and social sciences: from conceptual exploration to distinct ways of experimenting", *Journal of Artificial Societies and Social Simulation*, Vol. 13, No. 1, 5, 2010, http://jasss.soc.surrey.ac.uk/13/1/5.html; Varenne (F.), *Modéliser le social: méthodes fondatrices et évolutions récentes* [Modelling the social: founding methods and recent developments], Paris, Dunod, 2011, pp. 165–174; Varenne (F.) "Modèles et simulations dans l'enquête scientifique: variétés traditionnelles et mutations contemporaines" [Models and simulations in scientific investigation: traditional types and contemporary variations], in F. Varenne, M. Silberstein (Eds), *Modéliser & simuler: épistémologies et pratiques de la modélisation et de la simulation* [Modelling and simulating: Modelling and simulation epistemologies and practices], Vol. 1, Éditions Matériologiques, Paris, 2013, pp. 11–49; Varenne (F.) *Théories et modèles en sciences humaines. Le cas de la géographie* [Theories and models in the human sciences. The case of geography], Paris, Éditions Matériologiques, 2017, pp. 80–132.

3 Thus, in didactics, we speak of the *epistemic function of the written word* – in contrast to oral – to denote the way in which the written word facilitates the acquisition, fixation and diffusion of types of knowledge that have themselves already been represented in propositional terms.

4 Most epistemologists, even those most focused on the propositional format of knowledge, do not deny this.

5 See Hacking (I.), "Do we see through a microscope?", *Pacific Philosophical Quarterly*, 1981, Vol. 62, No. 4, 305–322.

6 With regard to the principles of models, see Varenne (F.), "Modèles et simulations dans l'enquête scientifique: variétés traditionnelles et mutations contemporaines" [Models and simulations in scientific investigation: traditional types and contemporary variations], 2013, op. cit., pp. 24–27; Varenne (F.) *Théories et modèles en sciences humaines* [Theories and models in human sciences], 2017, op. cit., pp. 110–112.

7 Laubichler (M.D.), Müller (G.B.) (Eds), *Modeling Biology: Structures, Behaviors, Evolution*, Cambridge, The MIT Press, 2007, p. 8.

8 The qualifier "heuristic" cannot, strictly speaking, be used to specify a first-degree epistemic function, but only a second-degree one. A model cannot be termed heuristic (i.e., facilitating the discovery or formulation of knowledge) in absolute terms. It is only heuristic relative to a desired form of knowledge: it is heuristic *for* observation, *for* explanation, *for* understanding, or else *for* theorization, etc. Furthermore, the cases provided in Laubichler and Müller's compilation merely confirm the secondary nature of their heuristic dimension.

9 One of the seminal epistemological publications defending this view is Morgan (M.S.), Morrison (M.) (Eds), *Models as Mediators*, Cambridge, Cambridge University Press, 1999.

10 Minksy (M.), "Matter, mind and models", in W.A. Kalenich (Ed.), *Proceedings of the International Federation for Information Processing (IFIP) Congress*, London, Macmillan, 1965, pp. 45–49; p. 45.

11 See, in particular, Margaret Morrison, "Models as autonomous agents", in Morgan (M.S.), Morrison (M.) (Eds), *Models as Mediators*, Cambridge, Cambridge University Press, 1999, pp. 38–65; Erika Mansnerus, "Explanatory and predictive functions of simulation modeling", in Gramelsberger (G.) (Ed.), *From Science to Computational Sciences*, Zürich, Diaphanes, 2011, pp. 177–193.

12 Black (M.), *Models and Metaphors*, Ithaca, Cornell University Press, 1962; Hesse (M.B.), *Models and Analogies in Science*, Notre Dame, University of Notre Dame Press, 1966.

13 It is not just because the model is autonomous, but also because it becomes the substitute questioned object, that it is distinguished from another type of mediating object: measurement instruments.

14 Morrison (M.), *Reconstructing Reality. Models, Mathematics, and Simulations*, Oxford, Oxford University Press, 2015, p. 20.

15 The details are given in Varenne (F.), *Théories et modèles en sciences humaines* [Theories and models in the human sciences], 2017, op. cit., pp. 80–132.

16 I would consider that the concepts here are clearly defined general ideas that can be rigorously denoted by symbols, and which are intended to be applicable in the real world, i.e., beyond just their symbolization space.

17 A first version of this table appeared in Varenne (F.), *Théorie et modèles en sciences humaines* [Theory and models in the human sciences], 2017, loc. cit., pp. 130–132. I would like to thank the publishers for allowing me to use this adapted and translated version here.

18 The examples given here are rarely pure from a functional point of view: they may be used to carry out several functions at once. But they are often typical of each function.

19 Regarding these models, see Frances Yates, *The Art of Memory*, London, Routledge, 1966.

20 I share Margaret Morrison's scepticism (see *Reconstructing Reality*, 2015, op. cit., p. 18) regarding the possibility of providing a general characterization not just for comprehension but also for explanation. In this instance, I have chosen to highlight only the characteristics that are most often encountered in the many analyses of these concepts: the causal link for explanation, uniting under one single representation for comprehension. In this way, by "explanation", I mean the intelligible representation (i.e., by concepts) of a system of interactions or a mechanism (elements + actions) that are assumed to be the cause of a phenomenon that affects the target system. By "comprehension", I mean a unifying conceptual representation that can be mobilized by an unassisted human mind. We understand a phenomenon that is composed of a variety of sub-phenomena when we can, by means of a single mental (mathematical or logical) operation, reconstruct the gist of the structure of that variety.

21 See preceding note.

22 I consider that what Van Fraassen (Van Fraassen (B.), *The Scientific Image*, Oxford, Clarendon Press, 1980) calls "empirical sub-structures" are in fact "models" in the case of experimental science (function No. 14), even if they are not models of theories (function No. 11).

23 Using the term adopted by Paul Humphreys in *Extending Ourselves*, Oxford, Oxford University Press 2004.

24 For their characterization, see Varenne (F.), "Modèles et simulations: variétés traditionnelles et mutations contemporaines" [Models and simulations: traditional varieties and contemporary mutations], in F. Varenne, M. Silberstein, *Modéliser & simuler: épistémologies et pratiques de la modélisation et de la simulation* [Modelling and simulating: epistemologies and practices of modelling and simulation], Volume 1, 2013, op. cit., pp. 42–44. See also the next section of this chapter. Since type 1 simulation models do not, however, preferentially fulfil a precise epistemic function, but, on the contrary, may at times fulfil one function, at times another, and at other times a whole set of functions simultaneously, they have not been included in this table since, for each function, only typical examples are generally listed.

25 Basing myself on several cases, including the exemplary case of the French word "*filtrat*" (i.e., "filtrate" in English), which denotes the results of filtration, I have elsewhere suggested using the French term "*simulat*" to denote the result of a simulation. See Varenne (F.), "La reconstruction phénoménologique par simulation: vers une épaisseur du *simulat*" [Phenomenological reconstruction by simulation: towards a depth of the *simulate*], in D. Parrochia and V. Tirloni, *Formes, systèmes et milieux techniques après Simondon* [Forms, systems and technical circles after Simondon], Lyon, Jacques André Editeur, 2012, pp. 107–123. I suggested that this term be translated into English as the fairly logical term, "simulate", which could therefore take its place alongside the existing term "filtrate", with both words also serving as verbs in English, as we know. However, since I am aware of the reticence that the English philosophical language may have towards new uses of words, I will refrain from doing so and will settle for the term proposed here.

26 By "symbol", I mean the same thing as Nelson Goodman: "'Symbol' is used here as a very general and colorless term. It covers letters, words, texts, pictures, diagrams, maps, models, and more, but carries no implication of the oblique or the occult", in Goodman (N.), *Languages of Art: An Approach to a Theory of Symbols*, Indianapolis, Bobbs-Merrill, 1976, p. xi. A symbol is an object that is accorded the property of referring. With regard to this property of referring, Goodman points out that: "'Reference' as I use it is a very general and primitive term, covering all sorts of symbolization, all cases of *standing for*" in "Routes of reference", *Critical Inquiry*, 1981, Vol. 8, No. 1, pp. 121–132; p. 121), Goodman's own emphasis.

27 Parker (W.), "Does matter really matter: computer simulations, experiments and materiality", *Synthese*, 2009, Vol. 169, No. 3, pp. 483–496; p. 486. Although Parker immediately qualifies her claim by stating that "neither all of the states that it occupies nor all of its properties at any given time must be assumed to represent states or properties of the target system" (ibid., note 6), by continuing to assert that a simulation must nonetheless necessarily have a sequential dimension – whether representative or not – she forgets that an r-simulation does not necessarily have an ordered aspect with a temporal deployment.

28 The concept of denotational hierarchy is introduced by Goodman (N.) in "Routes of reference", *Critical Inquiry*, 1981, Vol. 8, No. 1, pp. 121–132; p. 126.

29 See Phan (D.), Varenne (F.), "Agent-based models and simulations . . . ", 2010, op. cit. For further explanation and examples on this subject, see also Varenne (F.), "Framework for models & simulations with agents in regard to agent simulations in social sciences: emulation and simulation", in A. Muzy, D. Hill and B. Zeigler (Eds), *Modeling & Simulation of Evolutionary Agents in Virtual Worlds*, Clermont-Ferrand, Presses Universitaires Blaise Pascal, 2010, pp. 53–84; Varenne (F.), "Chains of reference in computer simulations", working paper, FMSH-WP-2013–51, 2013, 32 pages, https://halshs.archives-ouvertes.fr/halshs-00870463.

30 See an overview of this debate in Morrison (M.), *Reconstructing Reality*, 2015, op. cit., Chapter 6.

31 Morrison (M.), *Reconstructing Reality*, 2015, op. cit., p. 245.

32 Von Neumann (J.), "The role of the digital procedure in reducing the noise level" (1948), in A.H. Taub (Ed.), *John von Neumann – Collected Works*, Vol. V, London, Pergamon Press, 1962, pp. 295–296. See Varenne (F.), "What does a computer simulation prove?", in N. Giambiasi and C. Frydman (Eds), *Simulation in Industry*, Proceedings of the 13th European Simulation Symposium, SCS Europe, Ghent, 2001, pp. 549–554; p. 553. See also Varenne (F.), *Qu'est-ce que l'informatique?* [What is computer science?], Paris, Vrin, 2009, pp. 71–94.

33 See Varenne (F.), "Chains of reference in computer simulations", 2013, op. cit.

34 It is possible, nonetheless, to try to programme this emergence. See Varenne (F.), Chaigneau (P.), Petitot (J.), Doursat (R.), "Programming the emergence in morphogenetically architected complex systems", *Acta Biotheoretica*, 2015, Vol. 63, No. 3, pp. 295–308.

35 For the case of an analogical simulation (for example, a scale model of a boat in a rheology tank), the interactions between the parts of the test specimen correspond to

the "interactions between things or symbols of things" that I describe as taking place during the first step of any simulation.

36 This was the choice made by Wendy Parker (2009), as we have seen.

37 A sequence of mediations between the object of apprehension at the end of the sequence and the target system is itself a direct – global – mediation between these two objects. A mediator belonging to this type of sequence of mediations, however, is not a direct mediator between the apprehended object at the end of the sequence and the target system, but it is a mediator between intermediary objects.

38 There is, in effect, a possibility for simulation models of this type to provide – by means of their participation in this sequence of mediations – not just the functions of description and prediction, but also the function of explanation. To my mind, this seems to confirm the analyses proposed on this subject by Erika Mansnerus, 2011, op. cit.

39 Winsberg (E.), "Sanctioning models: the epistemology of simulation", *Science in Context*, 1999, Vol. 12, No. 2, pp. 275–292; Winsberg (E.), *Science in the Age of Computer Simulation*, Chicago, University of Chicago Press, 2010.

40 Winsberg, 2010, op. cit., p. 12.

41 Humphreys, 2004, op. cit., pp. 130–132.

42 Winsberg (E.), "Sanctioning models: the epistemology of simulation", op. cit.

43 Winsberg (E.), "Handshaking your way to the top: simulation at the nanoscale", *Philosophy of Science*, 2006, Vol. 73, pp. 582–594.

44 See Varenne (F.), "Chains of reference in computer simulations", 2013, op. cit.

45 See Winsberg (E.), "Reliability without truth", in E. Winsberg, 2010, op. cit., pp. 120–134; Lenhard (J.), "Artificial, false, and performing well", in G. Gramelsberger (Ed.), *From Science to Computational Sciences*, 2011, loc. cit., pp. 165–176.

46 Sismondo (S.), "Simulation as a new style of research", in G. Gramelsberger (Ed.), *From Science to Computational Sciences*, 2011, op. cit., pp. 151–163; p. 157.

47 See Carusi (A.), Hoel (A.S.), Webmoor (T.), Woolgar (S.), *Visualization in the Age of Computerization*, New York, Routledge, 2015.

48 See Varenne (F.), *Chains of Reference in Computer Simulations*, 2013, op. cit.

49 I will draw on examples presented in Chapters 1 to 7 and in Varenne (F.) *Du modèle à la simulation informatique* [From model to computer simulation], Paris, Vrin, 2007, as well as on older models described in the following works: Varenne (F.), "Le destin des formalismes: à propos de la forme des plantes: pratiques et épistémologies des modèles face à l'ordinateur" [The fate of formalisms: regarding plant forms – Practices and epistemologies of models in response to the computer], PhD, University of Lyon, 2004; Varenne (F.), *Formaliser le vivant: lois, théories, modèles?* [Formalizing living beings: laws, theories, models?], Paris, Hermann, 2010.

50 See Varenne (F.), 2004, op. cit., Chapter 1; *Formaliser le vivant*, 2010, op. cit., Chapter 3.

51 Varenne (F.), 2004, op. cit., Chapter 2; *Formaliser le vivant*, 2010, op. cit., Chapter 4.

52 Varenne (F.), 2004, op. cit., Chapter 7; *Formaliser le vivant*, 2010, op. cit., Chapter 9.

53 Varenne (F.), 2004, op. cit., Chapter 8; *Formaliser le vivant*, 2010, op. cit., Chapter 11.

54 Varenne (F.), 2004, op. cit., Chapter 9; *Formaliser le vivant*, 2010, op. cit., Chapter 12.

55 Varenne (F.), 2004, op. cit., Chapter 10; *Formaliser le vivant*, 2010, op. cit., Chapter 13.

56 Lenhard (J.), Winsberg (E.), "Holism, entrenchment, and the future of climate model pluralism", *Studies in History and Philosophy of Modern Physics*, 2010, Vol. 41, pp. 253–262.

Conclusion

Three different types of lesson may be drawn from the investigation set out in this volume. I will present a quick overview of each in this section. The first type of lesson touches on the methodology of philosophy of science. The second type is historical in nature and concerns the observations and assessments that can be made in connection with the recent history of science and applied science, in particular regarding the consequences of the computerization of the sciences. The third type of lesson is epistemological. It concerns, in particular, the issue of the nature of the knowledge that is produced by the new forms of integrative simulation. My aim is above all to shed what I hope will be new light on the thorny issue of the differences and affinities between knowledge through simulation and empirical knowledge. But I will also explore how a simulation's epistemic functions are enriched and complexified when the simulation is integrative; the epistemological meaning of the process of remathematizing simulations; and also the equally epistemological consequences of the new interdisciplinarity that is now permitted in contemporary science through the construction of these integrative virtual objects.

First of all, from the point of view of philosophical methodology, this research aims to demonstrate the importance of a philosophy of science approach that claims to be both empirical and theoretical. The primarily empirical part is built around an analysis of comparative, synchronic and diachronic cases, but it also contains occasional epistemological analyses. These analyses were necessary to explain the precise methodological choices of the players involved. In this extended and ramified case study I have sought to go beyond the specific case studies that are already available regarding modelling and simulation approaches in contemporary science. I wanted my study to make it possible to follow and explain not just a single school of modelling or a particular methodological choice, but rather the different technical, methodological and epistemological choices that the rise of computers has allowed to emerge but also has brought into conflict. I also wanted it to show how a new modelling solution and its accompanying epistemology were invented, in light of the technical and epistemological limitations of similar closely-linked alternative solutions. The most theoretical part of this book was presented mainly in Chapter 8. This part is the result of a number of inductions, as well as of corrections and updates to the original version of this comparative case study, which was first published, in

French only, in 2007. It is also the product of a large number of later collective and collaborative works, some of which I co-directed,[1] as well as being the fruit of other extended case studies that I carried out alone, primarily in the field of geography.[2]

The end of the new chapter (Chapter 8) in particular demonstrates that the path taken by these different comparative investigations may end up nowadays by showing – albeit very briefly here, and once again with reference only to the case of plants – the capacity for enlightenment, discriminating analysis and re-explanation imparted by the conceptual classifications and distinctions that were introduced. This empirical philosophy of science practice that I am trying to hone is based not only on comparative empirical analyses or on inductions in other words, but also on corrected conceptual analyses. Rather significantly, this practice has recently even taken a distinctly experimental turn, since its results are also incorporated in the new science in the making. Indeed, the results of this practice are used by modellers themselves, a number of whom have in fact quite extensively adopted the proposed conceptual classifications. They do so to explain the precise and differentiated specificities of the epistemological profiles of their own models and simulations, both to themselves and to their colleagues.[3] In return, such clarifications sometimes help them develop the drafts of new forms of modelling.

If we now consider the outcomes of this book in terms of the recent history of science and applied science, one of the main outcomes is as follows: it was indeed the emergence of the computer that was in large part the driver of the recent evolution of modes of formalization of complex systems such as the plant. Thanks to the computer, replications of forms that were already geometrically complete, and were simple but occupied a predefined space (triangles, squares, etc.), could once again become formalisms. A certain re-spatialization of formalisms became possible. It has been possible to re-inject form since the very moment when computation became automatic and when the local rules of reiteration could be taken into account without still requiring a formal condensation in the form of equations or mathematical models, strictly speaking. With the different types of transition from mathematical model to simulation, the computerization of science has thus fostered the deployment of formal models not only on the scale of time, but also in the dimensions of ordinary space. As far as the case of plants is concerned, as the formalisms became spatialized, it became possible to formalize the spatial forms in a more flexible and accurate manner. Since computer languages have become less and less bound by the abstractive formulations of traditional mathematics, the models that are implemented in these languages became first algorithmic, and then object-oriented, to the great benefit of semi-realistic simulations.

This was not yet enough, however, as the simulations that resulted bore little resemblance in their details and could not be calibrated on field data. Neither the internal heterogeneity nor the complex systematicity of plant shapes during growth could be grasped by the computer. Although the first simulations of plant morphogenesis were initially inspired by the paradigm of computation in physics and in logic, in reality they remained mere speculation for nearly twenty years before the concept and the technique of pluriformalization and fragmented modelling took root in an agronomic context. At the start of the 1970s, however, the Bingerville research station in Côte d'Ivoire was still far from benefiting from computational techniques like those that Murray Eden already had at his disposal in 1959 at MIT. Throughout the comparative history I have recounted, the transition from model to replicative simulation has in fact been revealed to be the result not just of the wider availability of computers, but also of a series of two major epistemological decisions regarding the epistemic functions of models of complex objects: 1) in order to predict a complex system, it was perhaps no longer preferable to analyse it, but rather to resynthesize it; and 2) in order to be able to correctly synthesize this complex system, it would be helpful to at least partially explain it on a micro-scale. It should be recalled that these two decisions did not initially arise from a desire to visualize the plant graphically, but rather to take a precisely quantified model and make it operational and predictive. In this regard, Philippe de Reffye benefited from a certain lack of prejudice: unlike his more epistemologically biased colleagues, he decided not to seek a monoformalized representation of the plant at all costs, even in a probabilistic form. He was also bolstered by a constant belief, admittedly somewhat ahead of his time, in the existence of "laws of nature" – although he accepted beforehand that they would have a complicated formal intelligibility. Because of the context and the particular field issues he faced, he was motivated primarily not by the mathematicism, computationalism or systemism that were common in the theoretical biology of the time, nor even by a physicalistic epistemology, but simply by the applicability of the models he had conceived. This was what led him to abandon statistical biometry and modelling, along with numerical simulations that were simply suggestive, heuristic or with theoretical aims, in order to usher the study and modelling of plant morphogenesis into the age of software- and object-based simulation.

Although mathematical models had sufficed for many years for other biological or ecological phenomena such as metabolic dynamics, matter and energy flows, or population dynamics, and replicative simulation did not appear to be a necessity, in the case of plant shapes the evolution of the epistemic status of mathematics and computers continued unabated for over forty years. Thus, in the case of plants, there was a transition first from theory to mathematical model and then from mathematical model to computer simulation. Indeed, looking at the plant as a whole, what stands out most sharply is what I called its highly composite nature: during its ontogenesis, a considerable spatial as well as temporal heterogeneity emerges for the rules of growth. No known mathematical model can grasp from the outset *both* this historicity and this spatial distribution. The plant is revealed as a distributed living being due to its highly populational nature. But its population differentiates over the course of time. Its ontogenesis has the advantage of substantial distribution and evolutive reiterations. Furthermore, its morphogenesis exhibits an altogether remarkable interaction with its environment that also precludes the use of purely physics-based theories when aiming to model these interactions. Lastly, although it is true that a growing plant seeks a certain morphological and functional optimum, it does so along paths that intermingle levels and causalities in a very muddled manner. Yet it was precisely the simulations' placement in space that made it possible to disentangle some of these paths, in order to better reflect the historicity of the whole. In fact, the transition to simulation has shown that neither the spatiality nor the historicity could be dispensed with; both must be taken at once. It was this preliminary openness in the formalization of such complexity that, unlike the rigidity of mathematical or physicalistic models, was very quickly grasped by several researchers working in the field of plant studies. For that matter, it is precisely this openness that seems to apply today to other disciplines – to the study of social morphogenesis, for example – as is shown by another case in point; the case of models of city growth.[4]

From a different, more epistemological point of view, we can see other results emerge from this study. First of all we see the fact that – in the case of plants, at any rate – the existence has been confirmed of a third source of knowledge, alongside theory and experiment, through the recent techniques of realistic and integrative simulation. Virtual testing in arboriculture and silviculture (the effects on shoots of bending under self-weight loading, cutting, pruning and thinning) is already operational and has been in regular use since the early 2000s. Based on Dauzat's work, simulations of physical and meteorological phenomena on plant simulations (simulations on simulations, or supra-simulations) have also been carried out in order to produce a simulated empirical signal for use in interpreting real satellite images. In 2006, following the integration of the physiological functioning of plants in the first models of plant architecture and structure simulations – once again with the help of computer implementation and its capacity for making formalisms converge – the "virtual agronomic experiments" forecast in 1995 finally became a reality. On this now reliable basis, major contracts representing sums of hundreds of thousands of Euros over a number of years were set up between the AMAP laboratory and beet producers, for example. In this way, the computer has finally become a "virtual laboratory" or, at the very least, the place where systematic virtual experiments in agronomy could be conducted, even though Prusinkiewicz's team had announced this new dawn too prematurely.

This historical and epistemological study has thus demonstrated that a computer simulation may be said to serve as a "virtual experiment". In this regard, the simulation no longer has just a simple difference of degree with a "conceptual argument", as Stephan Hartmann[5] and Manfred Stöckler[6] first suggested. It is thus not always a conceptual argument that has simply been transferred to a machine. It is important to emphasize this epistemological point, however, and take a further look into what characterizes computer simulation in all its variety. This investigation makes it possible, in fact, to draw three epistemological lessons on this issue.

First of all, not every simulation is necessarily a model immersed in time.[7] In 1951, when Ulam spatialized Von Neumann's self-replicating automaton model, he sought to immerse this first kinematic model in space, and not in time. The temporality of the calculation process was not decisive in this instance, especially because the calculation involved reiterations of multiplications of finite-rank matrices on an infinite space in order to simulate a stochastic iteration in a totally deterministic manner. This was the forebear of cellular automata, as they were subsequently named by Arthur Burks. But this forebear was deterministic, in contrast to the simplified and extended use that Murray Eden and many others rapidly made of it.

Furthermore, Hartmann pointed out that a simulation imitates a process by means of another process,[8] but this characterization seems too loose in this instance: unless the word "process" is given an unaccustomed meaning, the same could be said of a mathematical law such as the law of falling bodies, which – when simply calculated but not simulated on computer – can indicate, over a period of computation time that is iconically similar to the actual time of the target system, the fall height, z, reached as a function of time, t, using the simple relation $z = \frac{1}{2} g \cdot t^2$, where g is the acceleration constant at the Earth's surface (9.81 m/s^2). This type of calculation by computer is not a simulation, however. Lastly, I will not go into detail again on the updated distinction that I drew – using the contrast between the first AMAPsim software and the AMAPpara software as an example – between resemblance *by* dynamics and resemblance *of* dynamics. Not only is it not necessary that the computation process should correspond at each step of time to a realistic phase of an actual process, as Wendy Parker essentially acknowledged, but the result of a process-simulation, termed "result-simulation" or "r-simulation" in Chapter 8, does not necessarily need to have a temporal aspect.

Second, not every simulation is necessarily the calculation of a model. Here too, we must agree: if, by "model", we mean the mould or the container of all the heterogeneous rules that are set up in order to carry out a series of computations, then it is clear that there can be no computer simulation without a model. But, in this instance, we include something more restrictive in the term "model": the hypothesis of a formal homogeneity and internal consistency of the rules. From this point of view, there are simulations without a model, as Alain Franc, an INRA modeller and ecologist, already contended in 1996.[9] It is primarily the simulations that I called "pluriformalized", as well as agent-based simulations, that fall into this category. Thus, despite Paul Humphrey's words to this effect, it does not appear that a process of simulation of this sort gives rise solely to what he calls a "*computational template*".[10] The distinction I proposed in Chapter 8 between *model-simulation model* and *system-simulation model* may be instructive here. Admittedly, Humphreys clarifies certain points with his concept, but he continues to systematically give a unity and homogeneity that seem completely excessive to this *formal template*, including for the agent-based simulations that he also aims to include in this range.[11] It is this restrictive assumption that logically leads him to conclude that computer simulations in general only extend our powers of computation or perception in accordance with one particular aspect, which explains the "selective realism" that he ultimately advocates. Despite the realism that he defends – and which I am also ready to defend – he is not far from ultimately reducing simulation to the same selective nature as that which the classic mathematical models themselves often and explicitly possessed, long before computer simulation had been developed. This necessarily selective aspect of simulations appears to me today to be highly debatable and, at the very least, secondary. When we seek to represent a nuclear explosion almost in its entirety using the French Atomic Energy Commission's integrative computer simulation, we no longer do so just to acquire a certain focused view, or to investigate a particular sizing of fissile material, as was the case, on the contrary, of the numerical simulations in nuclear physics during the 1940s.

In light of these conceptual limitations, I came to consider that computer simulations should no longer be defined essentially on the basis of their dependence on a formal model. To me, they seemed, first and foremost and on a broader level, to be symbolization strategies and practices that might or might not have formal models as test specimens. In its procedural dimension, this symbolization practice took the form of a process of dynamic and distributed processing that was conceived with a view to a *partial and/or fictive replication* of elements, behaviours and/or global phenomena (e.g., history of states) or local phenomena (e.g., individual trajectories and interactions) followed by a *measurement* or a sensory rendering (graphic, audio, haptic) of that replication.

The important thing is that the simulation thus always makes a *sub-symbolic use* of certain traditional mathematic or logical symbols: these symbols remain symbols, of course, but they should also be considered as sub-symbols, since they *partially* lose their purely conventional nature in order to take on the character of exemplification. The symbols of certain simulations may at times lose this conventional nature from several aspects at once, as in the case of pluriformalizations, which explains the irreducible plurality of their perspectives. This exemplification character is of course relative, as I explained in Chapter 8. It may be a computer memory address, a pointer in the C++ programming language, or an object, in the sense of the object-based programming languages, that represents a neutron or molecule in a one-to-one relation. Although admittedly a simulation is not always itself a type of iconic model in which a single component of a system is represented by a single symbol of the simulation model, it nonetheless implements, in the broadest sense, a *dilated symbolization mode* from at least one aspect. Simulation goes against the function of *condensation* and abstraction of the traditional symbolic instruments of a logical and mathematical type. It is always a practice of partial demathematization followed by a series of interactions, and then by a form of measurement. This measurement practice may consist of an observation experiment carried out by the modeller through the use of a screen, for example. But it may also consist of an outright, en bloc, reuse and recruitment of the r-simulations in a supra-simulation.

Third, it may also be said that, under these conditions, there is not one but several types of computer simulation that we must be able to distinguish. There are even, in fact, at least two levels of distinction. The first level comprises distinction according to the form of sub-symbolization and demathematization that is primarily used. Thus, I distinguished between simulations that were discretized (Turing), spatialized (Ulam), probabilistic (Eden), geometrical (Cohen), or logical (Lindenmayer). Obviously these distinctions become more complicated once the sub-symbolizations are combined, as is often the case: in these instances, we may speak of simulation that is discretized-spatialized, discretized-stochastic or discretized-deterministic, logical-stochastic. Then there is a second level: distinction according to principle. This level was discussed in Chapter 8. To determine this we no longer ask: what is the nature of the result of sub-symbolization? What is the set of axioms or rules that acts as sub-symbolic for another set of axioms or rules? Instead we must ask: what must be done in order to reach this sub-symbolic formulation? It should be recalled that there are at least three ways to achieve this, which should be clearly differentiated.

1   Either *starting from available mathematical formulations* that are in effect *demathematized* by means of sub-symbolizations, according to one or several aspects.
2   Or *starting directly from algorithms* or from *rules*, in the case where, for example, there is no pre-existing theory or mathematical model.
3   Or *starting from formal objects that we standardize and formalize* without using either mathematics or a single logical formalism (for example in multi-agent systems) by relying on descriptive knowledge that is available elsewhere, as well as on metaphors that are available in non-theorized science or even in collective perceptions (common intuition, common meaning).[12] The fact that we rely on variable rules in order to formalize composite or heterogeneous objects in a variable and sometimes multiple manner (as in certain multi-scale simulations) means that this simulation is no longer primarily conceived in order to resolve calculability problems specific to mathematics, even though these problems retain their *regional* interest in these simulations, in particular in the later problems of *remathematization*.

These three types, which are described and exemplified in greater detail in Chapter 8, represent successively the numerical, algorithmic and software-based simulations. This sequence corresponds fairly well, however, to the way in which computer-simulation has evolved historically, as illustrated by the progress in the calculation powers of computers and in the expressivity of programming languages. The transition from model to simulation therefore cannot be reduced simply to a transition from laws to algorithms, even though this transition, which is already long-established in many sectors – as we saw, in fact, for the sector under study here – remains innovative in other sectors, such as in the new nanosciences, in particular.[13]

Turning now not to the identity,[14] but to the epistemic kinship between simulation and experiment, we find several reasons why this kinship may be invoked in light of these different distinctions. Without entering into over-long analyses here, a quick review citing at least four of these reasons can be drawn up. Of these four, only the first two have already been explained in publications on methodology and epistemology; the second two have been less systematically reported.[15]

Reason one: the empirical character according to effects. Simulation, understood solely as a *result* or set of results of a numerical or sensory (graphical, audio, etc.) nature, produces output phenomena that are qualitatively (e.g., evaluation by eye of graphical outputs) or quantitatively comparable to what would be obtained from field testing. This is the type of simulation usage for which only r-simulation is used, on the one hand, and for which, on the other hand, this r-simulation is used as a model with epistemic function No. 1 or No. 2. For this type of case, a quantitative comparison of different series of effects may even be effected in a statistical manner. In this way, statistical experiment design can be carried out on r-simulations resulting from a large number of repeated p-simulations, as long as one control variable in the p-simulation design can be changed each time the computation program is relaunched. At this point, we encounter what Galison calls the *pragmatic* argument in favour of the empirical nature of simulations, because – without concerning ourselves either with the realism of the models they are based on, or with these models' foundation in a tried and tested theory, provided that they are considered to be correctly calibrated – the same statistical data-analysis procedures as those that are applied to the results of actual experiments can be applied in practice to just the r-simulations that are considered to be the *results* of the process-simulations.

Reason two: empirical nature according to cause. The formal model or models that are processed in the computation of the numerical, logical, geometric, probabilistic, etc. simulation can be considered to be partially realistic on an elementary or intermediary scale that is specific to them. This is the case, for example, for the individually symbolized neutron in a Monte-Carlo simulation model in nuclear physics, for the individually symbolized molecule in an *ab initio* molecular simulation model in chemistry, for the individually and spatially symbolized algae branch in a logical model of a Lindenmayer-type simulation, or for the local rules of social agent behaviour in a multi-agent and multi-process system. This type of empirical character may appear in a simulation that is itself the result (the effect) of a *system-simulation model (SSM)* with a preferential scale on which some of its symbols, sub-models or algorithms take on forms or behaviours that refer almost iconically to a target system or its properties, and are to that end considered to be barely or only slightly conventional. This argument in favour of the empirical nature of a simulation is often backed by a conception that Galison describes as essentialist, such as stochasticism,[16] for example, which holds that the simulated random event of a Monte-Carlo simulation represents an actual random event. During the 1960s, the geneticist Motoo Kimura considered that this type of simulation could be used, like an authentic *experiment*, to disprove theoretical mathematical models.[17] Margaret Morrison also advanced this argument once again with her recent focus on particle methods. In these simulation methods, "the particles in the simulation model can be directly identified with physical objects".[18]

Reason three: empirical character according to the intertwining of causes. There is a third argument, relating to an intermediary point between the causes and effects of simulation, which to my mind allows us to speak of virtual experimentation. This argument concerns the *pluriformalized simulations* in which there are several scales or perspectives that are simultaneously symbolized. Since they are "pluri-perspective" – in other words they do not favour the realism or validity of the symbols or rules of one given sub-model to the detriment of those of another – these simulations do not produce minimal models with one single unilateral interpretation. By symbolizing various constituent phenomena, and doing so on different scales, these simulations do not serve primarily as ad hoc tests of theoretical hypotheses, even though they may also carry out this type of function (No. 2) that more traditional computer simulations have been fulfilling for some time. Their own empirical nature comes partly from their different ways of capturing and harnessing field data in the various realistic and partially iconic sub-models that they intertwine (see reason two: empirical nature according to cause). But their empirical nature also comes from the evolutive and non-uniform nature, over the course of the computation, of the formal step-by-step rules of management of the relationships between these same formal sub-models. These rules may also be more or less realistic or fictive. Lastly, the management by events and not by clock of many pluriformalized simulations adds to the formally destructured and un-conceptualizable character of the sub-models' modes of coexistence and interaction. One form of historicity is represented here by the particular irreversibility caused by a reliance on earlier results not just on the part of some quantitative results of the subsequent computation, but also on the part of the actual qualitative type of the next rule to apply or of the next computation to be carried out so as to obtain the subsequent result. In my view, these two latter characteristics appear unprecedented and could explain the distinctive novelty of this third empirical-nature characteristic – a novelty that is specifically fostered by computerization and by the computer implementation of object-based system-simulation models.

Several authors, admittedly, have emphasized the fact that the manipulation and calculation of a mathematical model may bring up a few "surprises" and that, to that end, they can already be recognized as having a kinship with experiment.[19] Of course, this is even more the case for computer simulations. Mary Morgan objected, however, that the manipulation of a formal model could not, on the contrary, go so far as to confound the experimenter, i.e., leave them without the means of conceptualizing what they see or measure in output, since they may always, at least in principle, conceptualize, trace back and re-explain the result in the initial terms of the formal model that was imposed at the outset: "however unexpected the model outcomes, they can be traced back to, and re-explained in terms of, the model".[20] It is precisely this ability to conceptually trace back and re-explain the p-simulation, even if only in principle, that in my opinion should be disputed for pluriformalized simulation models. Certain system-simulation models of this type are not just surprising; they are also confounding. Thus it would be wrong to believe that it is possible – even in principle – to conceptualize and thereby depict in an abbreviated manner the entire process of computation to which they give rise. In reality, such computational simulation replaces the traditional formalisms by integrating several formalisms step by step, and making them computationally compatible. This has the effect, however, of producing an *aggregating conver-gence* between formalisms and not the *absorbing convergence*[21] that is usually practiced, on the contrary, by mathematics and associated theoretical disciplines, such as theoretical biology. This computational compatibility is sub-symbolic in nature: there is no known unified, abbreviating and general rule of genesis that could – even a posteriori – determine a homogeneous symbolic operation, in one single formal language. It should be noted that this is not an issue of algorithmic incompressibility, because even for the implementation of this complex type of simulation model, the executable version of the underlying computer program is of course generally algorithmically compressible. However, it is an issue of conceptual and cognitive incompressibility that holds for a different scale than that of the machine language or of the executable program. It applies on the scales of the terms of the initial sub-models, of those models that are formally conceptualized and symbolized at the start, of course, but that become computationally intertwined during the computation.

With the emphasis on this third empirical character, we get a better understanding of one of the major methodological and epistemological innovations that accompany the computerization of science and, in particular, the unstoppable transition from mathematical models to computer simulations: namely that the representable result of pluriformalized simulations cannot be immediately or directly homogeneous with a writing or with a set of exclusively propositional knowledge formats. The conceptual re-explanation of the result of these simulations is therefore not only impossible de facto; it is also impossible *de jure*. This is because these simulations do not fall under one single set of axioms nor one single grammar, whether formal or generative. Counter to a whole range of uses of formalisms in the empirical sciences, simulation becomes a process of formalization that no longer aims at symbolic condensation, but instead sometimes aims first at a symbolic dilation, which in turn enables the coexistence of several forms of knowledge – namely not just traditional formalisms but also non-propositional forms of knowledge – on levels where sub-symbols prevail. It is precisely these r-simulations, once they have been calibrated on the different scales,[22] that may once again serve as a virtual experimental ground and as test specimens for a process of supra-simulation. This was the case, for example, in Jean Dauzat's work.

Reason four: empirical character through lack of an a priori specific epistemic function. There are cases where an empirical dimension is conferred on a simulation by default because it is impossible to decide whether it is epistemically equivalent to a theory, a model, an explanation, a conceptual argument or an experiment. As a result, this default kinship is chosen in the first place because an experiment is often considered to be less informative, i.e., it provides less authorization for induction than all the other sources of knowledge. This nature, which also seems to appear in such simulations, is due to the fact that, along with the *axiomatic and conceptual heterogeneity* that I described earlier, there may also be *a heterogeneity of epistemic functions* specific to each sub-model. In the pluriformalized simulations that have been studied here, for example, sub-models with descriptive epistemic function No. 7 (stochastic branching models, phenomenological mathematical models, digitizations of actual scenes) follow *step by step* in the computation alongside sub-models with an at least partially explanatory function No. 8 (mechanical branch-bending model, models of local effects of shade on photosynthesis, model of local mechanisms of the creation and then the flow and allocation of matter in the plant). This phenomenon was mentioned quite early on and fairly often by several authors, without ever being truly clarified, to my mind. Many authors prefer to resort to a suggestive metaphor of the type chosen – aptly, it must be said – by Marcel Boumans, reflecting the heterogeneity of ingredients used in cooking: "Model building is like baking a cake without a recipe. The ingredients are theoretical ideas, policy views, mathematisations of the cycle [in economics], metaphors and empirical facts".[23]

Let us try briefly here not to resolve this difficult problem, but to define its nature a little more precisely. First of all, it can be seen from the earlier discussion that simulations *intertwine sub-models, step by step, with distinct epistemic functions* that do not, themselves, have a simple resulting epistemic function. For such simulation, in fact, there is no a priori law of internal composition that is applicable in most cases for the different epistemic functions of the sub-models that produce it. Furthermore, it should be noted that the epistemic function of such complex simulation models no longer depends simply on the uses nor on the particular perspective of the user alone, as was the case, on the contrary, in the classic practice of monoformalized mathematical models. Nor does this epistemic function stem simply from the method of construction of the simulation program alone. In actual fact, this function is to a great extent determined by the different types and degrees of validation of the sub-models involved, where these types and degrees themselves hold for the different epistemic functions that these sub-models provide first separately, and then together. This is why, when such a global and integrative simulation model is not completely explanatory, it cannot be said to be solely descriptive either. The alternatives are no longer as simple. Indeed, such a simulation model simultaneously provides several epistemic functions, which until recently were often considered contradictory. It should be understood, however, that the simulation model does not provide these functions for the same scales or for the same aspects. This type of simulation model can, for example, be used for indirect identification of the parameters of a descriptive or explanatory sub-model, which suggests that the resulting simulation, in turn, presents a good degree of validation on a global scale for function No. 7 (descriptive), at the very least, as well as on several other intermediary scales for its already calibrated sub-models, with function No. 7 or perhaps even with an aspiration to function No. 8 for some of the sub-models that are newly expected to partially explain local phenomena, for example.

Where does the reliability of such complex simulation models come from? As can be seen, it probably does not come solely from an almost-automatic process of optimization along the lines of selection by trial and error. I would say that it is precisely this multiplication – not just of data but also of the epistemic functions of the sub-models and of their types of correlative basis, which at times is anchored in theories and at other times in forms of data of different types – that increases the progressive reliability of the inductions that may be drawn from such simulation models.[24] This composite nature may be seen either positively or negatively. In some cases, no doubt, as Lenhard and Winsberg saw, it may be considered as an entrenchment, and thus as sign of a lack of specific legitimation and as indicative of an uncontrolled opacity and rigidity.[25] For others, however, it must be seen on the contrary as indicative of open-mindedness, intellectual humility and concern for heightened realism. It may be the surest sign that modellers have left behind the illusions of a desire for a unified mathematical theory, since that theory may turn out to be premature or vain in certain areas relating to life sciences or social science. Such a global simulation, once it has been calibrated and stabilized, may in certain cases replace the target system in order to be used for the identification of certain sub-models applicable on a scale that is inaccessible in the field. Many simulations rely on this type of indirect identification, and not only in the life sciences; this is the case, for example, in astrophysics, for the indirect observation of planets or black holes that are not directly observed in reality.

Even though these practices may be convincing, the precise foundation of the legitimacy of the almost-empirical type of information on which they rely is very difficult to grasp. Indeed, it is the obscurity of this foundation that is at the root of many disagreements and counter-arguments between philosophers. This obscurity is itself, to my mind, closely linked to the complexity of the *hierarchical and inter-determination relations* between the epistemic functions of the different sub-models. This is because, in the final analysis, the sub-models' indirect identification practices and the resulting indirect induction of representations of actual phenomena are in fact based on the hypothesis of the existence of an *epistemic hierarchy* between the epistemic functions of the global simulation and the epistemic functions of the other sub-models that are computationally intertwined. These hierarchies and orders of mutual determination, however, are only very rarely completely elucidated and justified by the scientists themselves. A considerable work of epistemological analysis therefore still remains to be done.

The fact remains, of course, that this particular indirect sub-model identification practice also introduces problems of underdetermination. These problems, however, may seem troublesome precisely only when a direct informative use is intended to be made of this identification of sub-models. This practice itself can only have a phenomenological and transitory aim, however. This is the case when trying to obtain a calibrated integrative simulation whose implementation can then be used as an empirical test ground for remathematizations, i.e., for the conception of mathematical models that are more workable and less problematically optimizable than a simulation model, in particular with a view to improving farming or forestry crop management sequences, for example. This is the direction that de Reffye and his team have taken in their work since 1998.[26]

To conclude, I would say that the historical and epistemological analysis that I have presented here may also be seen as a contribution to the debate on the most favourable collective working methods for interdisciplinary research, since this debate involves dealing jointly with complex objects and sets of problems. What this study shows in no uncertain terms is that agreeing verbally on an objective, engaging in dialogue in this respect, even going so far as to adopt a common language or a sort of shared lingua franca or Creole, all these language practices are still far from being sufficient to effectively agree, i.e., in an effectively operational way, about the complexity of systems such as living systems. It is necessary to go all the way to a *common object*, and not just as far as a common language or aim. This is undoubtedly the underlying reason why replicative and controlled virtualization, i.e., integrative and figural simulation, has ultimately established itself among plant specialists, be they physiologists, ecophysiologists, botanists, agronomists, forestry experts, or population and plant-population ecologists, in the same way as it has established itself in the integrative brain-simulation models, and will no doubt soon establish itself more firmly in the human sciences that use modelling. This is in fact one of the aspects of the computerization of science that allows this kind of robust composite production. *Only when researchers are faced with the same object and not just in the same language or the same system of signs, even if this is a Creole, can they truly progress together.* In order to achieve an interdisciplinary scientific research that aims to reach a consistent operational practice and that refuses to stop at practices that are rhetorical, political or at practices for managing only collective representations, we cannot dispense with a *common referent.* Admittedly, this referent is virtual, secondary and constructed, but it is effective, and can be manipulated, modelled, projected, analysed and, if need be, remathematized. For the moment, it also appears to be the only *objective foundation* on which it is possible to construct and operate a *scientific interdisciplinarity that is both effective and controllable.* In this respect, the computerization of modelling practices has made such an advanced and unprecedented interdisciplinarity practice possible. We are now dealing with an interdisciplinarity of integration of models, and not just of cohabitation and juxtaposition of models. This interdisciplinarity is admittedly far from eliminating all the disagreements on scale and on points of view, but it has the merit of being flexible by first of all of reducing conflict between disciplines – a conflict that is at times their main component. By placing itself provisionally *below* all the dialectics that are ultimately cursory since they are valid on only one scale, this interdisciplinarity offers a mutual integration of competing conceptions that is both judicious and revisable. It is in this respect that simulation is incontestably fertile in the science of complex and composite objects: it offers a renewed space for interaction between forms of knowledge in which the barriers between disciplines do not disappear, but where they evolve more freely by taking on other meanings.

Thus, ultimately, we can say: the computer does indeed contribute to a redistribution of disciplines, to a re-establishment of their relationships and to a controllable blending of models, as well as to an effective integration of approaches. It is a movement of new integrations between disciplines that until recently were far removed, as we have seen here in the case of agronomy, botany, forestry and physiology. The exact limits of these integrations and convergences remains to be defined, but at the same time it is a movement of creation of new divergences, as shown by the astonishing revival of a research programme in theoretical biology at the very heart of the AMAP laboratory, and subsequently on its periphery. In its most innovative contribution, nonetheless, although it doesn't actually console us for the divorce from the real that it is orchestrating behind the new mediations and virtualizations that it permits, the computer also seems to contribute in return to the co-construction of a *common sense of the second type.* This is a co-constructed common sense, subject to a secondary *theoria* that must remain collectively revisable, precisely because it has been collectively formed.

All these practices for the construction of integrative models are admittedly not always conclusive in a visible or quantifiable manner. The final benefit of this investigation is that it demonstrates that such a practice, when it is sufficiently stabilized and mature, makes it possible to decisively capitalize on scientific knowledge. Thus, it should be recalled that remathematization is not the result of a remorse that apparently consumed plant simulators, who therefore suddenly returned to producing purely mathematical growth models. Instead, it was through deepening the epistemic demands that emerged from the intensive use of the first integrative and pluriformalized simulations that the simulators chose to remathematize. Furthermore, it was not the field data or the data models that they remathematized, but the p-simulation itself. This is proof that the remathematization phase could not have forgone the earlier phase of integrative simulation. In my opinion, it was indeed this necessary technical sequence that many of the disciplines specializing in complex, living or social systems began to put into effect: they no longer launched themselves into a search for a general theory – even with a local object – without first providing themselves with a simulation and thus with a multi-scale and multi-process capture of the complexity of the ground – a capture whose epistemic function is multiple, and both partially descriptive and partially explanatory. This in no way represents the end of theory in science, but simply the end of the naïve belief that it would always be enough to directly confront the data, models of data or empirical laws with mathematics in order for a testable and refutable theory to suddenly, and without any other formal mediation, spring to mind.

Notes

1 Varenne (F.), M. Silberstein (M.) (Eds), *Modéliser & simuler: épistémologies et pratiques de la modélisation et de la simulation, Volume 1* [Modelling and simulating: epistemologies and practices of modelling and simulation], Paris, Éditions Matériologiques, 2013; Varenne (F.), Silberstein (M.), Dutreuil (S.), Huneman (P.) (Eds), *Modéliser & simuler: épistémologies et pratiques de la modélisation et de la simulation Volume 2* [Modelling and simulating: Epistemologies and practices of modelling and simulation], Paris, Éditions Matériologiques, 2014.

2 Varenne (F.), *Théories et modèles en sciences humaines. Le cas de la géographie* [Theories and models in human sciences. The case of geography], Paris, Éditions Matériologiques, 2017.

3 For a reprise of the concept of remathematization in plant modelling, see, for example, Cournède (P.-H.), "Dynamic System of Plant Growth", Habilitation Doctorate, University of Montpellier, 2009, pp. 7–8, p. 47. For a reprise of the twenty-one functions of models in a sociological context, see, for example, Forget (A.), "La modélisation" [Modelling] *in* B. Gauthier, I. Bourgeois (Eds), *Recherche sociale – De la problématique à la collecte des données* [Social research – From problems to data collection], 6th edition, Québec, Presses de l'Université du Québec, 2016, Chapter 6, pp. 129–158. For a reprise in geography, see Tannier (C.), "Analyse et simulation de la concentration et de la dispersion des implantations humaines de l'échelle micro-locale à l'échelle régionale: modèles multi-échelles et trans-échelles" [Analysis and simulation of the concentration and dispersion of human settlements from a micro-local to a regional scale: multi-scale and trans-scale models], Habilitation Doctorate, Besançon, University of Burgundy, 2017, pp. 10–16; 107–112.

4 Batty (M.), *Cities and Complexity*, Cambridge, The MIT Press, 2005; Phan (D.), Amblard (F.), *Agent-based Modelling and Simulation in the Social and Human Sciences*, Oxford, The Bardwell Press, 2007.

5 Hartmann (S.), "Simulation", *Enzyklopädie Philosophie und Wissenschaftstheorie, Vol. 3* [Encyclopaedia of Philosophy and Philosophy of Science], Stuttgart, Verlag Metzler, 1995, pp. 807–809.

6 Stöckler (M.), "On modeling and simulations as instruments for the study of complex systems", in M. Carrier, G.J. Massey, L. Ruetsche (Eds), *Science at Century's End*, Pittsburgh, University of Pittsburgh Press, 2000, pp. 355–373.

7 Hartmann (S.), "The world as a process: simulation in the natural and social sciences", in R. Hegselmann, U. Muller, K. Troitzsch (Eds), *Modelling and Simulation in the Social Sciences from the Philosophy of Science Point of View*, Dordrecht, Kluwer Academic, pp. 77–100; Humphreys (P.), *Extending Ourselves: Computational Science, Empiricism and Scientific Method*, Oxford University Press, 2004.

8 He was followed on this issue by Parker (W.), "Does matter really matter: computer simulations, experiments and materiality", *Synthese*, 2009, Vol. 169, No. 3, pp. 483–496.

9 See Alain Franc's short text in Blasco (F.), *Tendances nouvelles en modélisation pour l'environnement* [New trends in modelling for the environment], Paris, Elsevier, 1997, p. 322.

10 Humphreys (P.), 2004, op. cit., pp. 130–132.

11 Since that time, however, Humphreys has recognized that his 2004 book underestimated the specificity of agent-based simulations. See Humphreys (P.), "The philosophical novelty of computer simulation methods", *Synthese*. 2009, Vol. 169, No. 3, pp. 615–626; p. 619. But, although I am prepared to accept most of his more recent claims about the novelty of computer simulation, I question whether he uses sufficiently powerful arguments against the resurgent structuralist theses because, even though he identifies two different roles played by time in a simulation, he does not clearly distinguish between simulation of a system and simulation of a model, or between simulation as a process and simulation as a result. Above all, he recognizes neither the fact that a simulation is not necessarily based on a unique formal model, nor that it is not necessarily designed to deal with propositional knowledge only. As a consequence, the new epistemic opacities of simulations essentially rely, according to Humphreys, on time and on the correlative computational complexity measures. This argumentative basis is likely to remain very weak, however, when seen from purely mathematical and structuralist points of view on computer simulations.

12 See Livet (P.), "Towards an epistemology of multi-agent simulation in social sciences", in Phan (D.), Amblard (F.), 2007, op. cit., pp. 169–193.

13 Lenhard (J.), "Nanoscience and the Janus-faced character of simulations", in D. Baird, A. Nordmann, J. Schummer (Eds), *Discovering the Nanoscale*, Amsterdam, IOS Press, 2004, pp. 93–100.

14 The point here is to question the kinship and not the identity, since, for that matter, I am convinced that, because a simulation does not involve causal relations of the same nature, it cannot itself be of exactly the same nature as the experience with the real world that it is simulating.

15 A first version of this analysis of the empirical nature of simulations was published in Varenne (F.), *Du modèle à la simulation informatique* [From model to computer simulation], Paris, Vrin, 2007, pp. 188–192. It also appeared in Phan (D.), Varenne (F.), "Agent-based models and simulations in economics and social sciences: from conceptual exploration to distinct ways of experimenting", *Proceedings of the 3rd edition of the EPOS Symposium (Epistemological Perspectives On Simulations)*, Lisbon, ISCTE, 2008, pp. 51–69. An updated and amplified version of this article appeared under the same title in Phan (D.), Varenne (F.), *Journal of Artificial Societies and Social Simulation*, 2010, Vol. 13, No. 1, 5, http://jasss.soc.surrey.ac.uk/13/1/5.html.

16 See Glossary.

17 See Dietrich (M.R.), "Monte-Carlo experiments and the defense of diffusion models in molecular *population genetics*", *Biology and Philosophy*, 1996, Vol. 11, No. 3, July, 339–356. Kimura was one of the first in the field of population genetics to represent genes in an individualized manner in individual-based system-simulation models.

18 Morrison (M.), *Reconstructing Reality*, Oxford, Oxford University Press, 2015, p. 221. Morrison's arguments in favour of the empirical nature of the knowledge obtained through simulation vary between an emphasis on what she calls "the identification with the physical system" permitted by particle methods (ibid., p. 222), which I have called their iconicity, and an emphasis on the fact that experiments – like simulations – especially because they

always require instruments and therefore models of apparatus, ultimately only give access to the target system by means of numerous mediations, of which several are models (ibid., pp. 220–221; p. 245). However, since, according to Morrison, all simulations are experiments on models – which I have demonstrated to be inaccurate – and all experiments are experiments through at least one model – which is, by contrast, acceptable – it would appear that, for Morrison, there is only a difference of degree, rather than a difference in nature between the empirical character of an experiment and that of a simulation. To my mind, quite aside from this fluctuating and debatable characterization of the meaning of the term simulation, there are also many other distinctions to be made. For example, a difference should be made between instrument and model, between instrument model and target-system model, between model simulation and simulation model, and last but not least between analysis model and synthesis model. Once this has been achieved, then a precise differentiation should be made between the sequence of models that is shown by an experiment and the sequence shown by a simulation, and their different modes of validation, both intermediary and global, must then be precisely evaluated and compared. Only then might we be able to explain exactly where the difference lies between the two empirical natures in question. This is not the place to undertake such work, however.

19 See, for example, Morgan (M.S.), "Experiments versus models: new phenomena, inference and surprise", *Journal of Economic Methodology*, 2005, Vol. 12, No. 2, 317–329.
20 Ibid., p. 324.
21 See Glossary.
22 This is aspired to but rarely achieved in human sciences today, unlike in botany – which explains the often speculative use of agent-based simulations in these sciences still today. This was also the case, however, of biologist Dan Cohen's simulations in their day.
23 Boumans (M.), "Built-in justification" in Morgan (M.S.), Morrison (M.) (Eds), *Models as Mediators*, Cambridge, Cambridge University Press, 1999, pp. 66–96; p. 67.
24 By virtue of this (only reasonable) empirical rule that determines that a representation may be considered more reliable and realistic a priori if it is based on a multiplication, crossing and cross-checking of access modes to the target system that are both different and independent. Regarding the simple idea that a multiplicity of independent sources – independent, that is, not only in terms of content, but also in terms of formats and forms of content – is still, in the final analysis, a major criterion of any scientific realism, even in the sophisticated types of realism found in contemporary philosophy of science, see Varenne (F.), *Théorie, réalité, modèle* [Theory, reality, model], Paris, Matériologiques, 2012. Somewhat similar ideas can be found in Wimsatt (W.C.), *Re-Engineering Philosophy for Limited Beings*, Cambridge, Harvard University Press, 2007.
25 Lenhard (J.), Winsberg (E.), "Holism, entrenchment, and the future of climate model pluralism", *Studies in History and Philosophy of Modern Physics*, 2010, Vol. 41, No. 3, pp. 253–262.
26 See their recent synthesizing publication: Reffye (de) (P.), Jaeger (M.), Barthélémy (D.), Houllier (F.) (Eds), *Architecture et croissance des plantes. Modélisation et application* [Plant architecture and growth. Modelling and applications], Paris, Quae, 2016.

Glossary

Notice:

Apart from the new entry under "Epistemic function", this glossary adheres to the technical terms included in the original French version of this book. In point of fact, although this new version, with its entirely new Chapter 8, proposes a whole set of other terms that complement the original glossary terms, it does so by introducing them from the outset in a definitional and classificatory way, along with illustrative examples. This holds true, for example, in the case of concepts such as "model nature", "model principle", "general function of a model", "main function of a model", "specific function of a model", "model simulation" or "simulation model". Thus, for these specific concepts, as well as for the twenty-one specific functions of models, and also for more in-depth definitions concerning the three types of computer simulation, the reader is advised to turn directly to Chapter 8 and its different sub-sections. For this reason, it seemed unnecessary to me to reproduce these definitions in the glossary, since they are already presented in a clear and easily identifiable manner in that chapter.

Absorbing convergence   An expression I introduced to describe a convergence between modelling formalisms at the end of which, due to the properties of intra-mathematical generalization (algebraic topology, theory of categories, etc.), one formalism is used to emulate and then reduce all the others.

Aggregating convergence   An expression I introduced to describe a convergence of formalisms, or of different formal perspectives on a modelled object, that occurs through step-by-step aggregation and intertwining during computation. This type of convergence takes place without prior mutual reduction (unlike absorbing convergence), and most often operates on a computational level (implementation) rather than a mathematical level. See the initial work of Philippe de Reffye and AMAP in this regard (1976–1996). See also "object-oriented programming" and "pluriformalization".

Algorithmic simulation   A type of computer simulation (in the restricted sense of an assisted formal calculation) of a discrete and stereotyped logical model. The first L-systems (1968) are thus highly standardized formal rewriting grammars and are therefore simply recursive. The simulation is no longer numerical in this context and substitutes the constant and consistent logical rules (an algorithm in this sense) for the step-by-step approximated processing of a mathematical law: there is a transition from the calculation of laws to the calculation of rules. This type of simulation is distinct from numerical simulation, and should also be distinguished from software-based and object-based simulations. It should be noted that an overly logicist epistemological view of computer simulations makes it impossible to draw the latter conceptual distinction. But this distinction is crucial in order to understand this essential part (a majority, since the end of the 1990s) of the sciences and technologies that use simulation, without however reducing their objects to homogeneous formal representations.

Allometry   See "mathematical model".

Architectural model   According to botanists Francis Hallé and Roelof A.A. Oldeman (1970), an "architectural model" of a plant characterizes its statistical habit and morphogenetic history as dictated by genetics for the first few years of the plant's life. It is based on four characteristics: 1) the type of growth (rhythmic or continuous); 2) the branching structure (presence or absence of vegetative branching; sympodial or monopodial branching; rhythmic, continuous or diffuse branching); 3) the morphological differentiation of the axes (orthotropic or plagiotropic); and 4) the sexuality position (terminal or lateral). An architectural model is determined when there is a particular combination of these morphological characteristics and their associated graphical symbols. Hallé and Oldeman identified 24 such combinations.

Biological stochasticism   Peter Galison uses the term "stochasticism" with reference to an ontological thesis according to which the world is discrete and "governed" by elementary stochastic events. This option evolved, in particular, in order to justify – sometimes a posteriori – the use of the Monte-Carlo method in nuclear physics, as well as in chemistry. By adding the qualifier "biological", it is extended here to the application of this same method to the cellular granularity of living tissues (see Murray Eden, 1960).

Biometry   Discipline established by Francis Galton, the initial aim of which was to apply statistical methods to the measurement of genetic traits, with a view to confirming Darwinian theory. Today, biometry includes all the quantitative methods applied to the measurement of living beings.

Computer simulation   Any formalization strategy that takes the form of a step-by-step and spread-out (or dilated) process of dynamic processing of partial replication of global or local elements, behaviours or phenomena (history of states) followed by measuring and/or rendering perceptible (by means of graphical visualizations, etc.) that replication. The simulation thus makes a sub-symbolic use of certain traditional mathematical or logical symbols: these symbols are once again reified and *partially* lose their conventional symbolic nature in order to regain an iconic nature (a computer memory address = a neutron or a molecule, etc.). Although it is not always an iconic model, a simulation is, in the most general sense, a *dilated representation* according to at least one dimension or aspect. It runs counter to the *conden-sation* function of traditional symbolic instruments (logic, mathematics). In this regard, a simulation is not always simply either the immersion of a model in time (as Stephan Hartmann suggests: doing so would restrict it to one single dimension of dilation), or the replication of a dynamic process per se. Although some of the simulation's results tend to be mimetic, it does not nec-essarily also have a mimetic dynamic. In other words, its successive states do not necessarily correspond mimetically to the states of the phenomena being simulated (this is the case of the first version of the AMAPsim software, compared with AMAPpara). For more distinctions and examples regarding computer simulations, see Chapter 8.

Discrete event simulation   A type of software-based simulation fostered by object-oriented programming, in which not only the numeric variables describing the system are discrete but also all the elements and possible events in the system are directly and strictly represented as discrete. Since the events are discrete, the time in the target system can also be directly rep-resented as discrete, and no longer as an *ex post* discretized representation of a previously given continuous time. As a consequence, the simulation model can directly represent the system as evolving by a process of jumps from one event to another. Time can also be seen as produced by events, rather than the contrary as was previously the case. Several types of time management (by clock or by event) may thus be chosen. Simulated time and calculation time may also be different, making it possible to manage the simultaneities. This type of simulation reveals the contingency in computer simulations of the hitherto frequent but unnecessary resemblance – or iconic relation of reference – between the time taken for computation and the time taken by the processes in the target system.

Epistemic function   Function of an object (which may be tangible, symbolic or mental) used in a scientific investigation. It not only denotes the function of guarantee, or of aid to accreditation or diffusion that this object appears to per-form *for* an already-formed *propositional* knowledge such as a belief, but also the *form itself of that knowledge* to which that thing allows access. In Chapter 8 I present twenty-one different epistemic functions for scientific models.

Experimental design   Notion popularized by the multivariate statistical analysis method introduced in agronomy by R.A. Fisher in 1919. An experimental design typically involves defining the goal of the experiment (identifying the influence of certain factors, etc.), defining the controlled and uncontrolled factors, describ-ing the types of measurements to be carried out, indicating the experimental units to be measured, giving a description of the way in which these units will then be taken into account in the statistical processing (with randomization if necessary, i.e., inclusion of a random event in the processing order) and, lastly, indicating the data analysis techniques (generally variance reduction).

Formal model (in the non-formal sciences)   Type of formal construct (of a logical or mathematical type) with a certain unity and homogeneity, such that it makes it possible to answer certain questions or fulfil certain functions (cognitive, empirical, communicational functions) regarding an object, a system or an observable phenomenon or situation. A formal model differs from theory because of its locality-linked nature, its prior adaptation to the initial questions asked and its inability to directly produce results of a theorematic type. It differs from computer simulation because of its native perspectivism, its enduring basis in easily mobilizable and manipulable symbolic systems (at least *de jure*) for a person without the help of a computer, and also because of the frequent and correlative requirement of formal homogeneity. See Chapter 8 in its entirety for further distinctions.

Fragmented modelling An expression I introduced to indicate a method that carries out, in a fragmented way, the modelling of a composite object such as a plant. From 1976, Philippe de Reffye separately modelled each law of apparition of growth or death of each type of coffee-plant organ, and required the computer program simply to manage this formal fragmentation by integrating these fragmentary sub-models into a simulation (or into a history of successive states of growth), proceeding step by step and bud by bud (from 1979). See "pluriformalization".

Geometric simulation In this context, indicates a type of plant-growth simulation aimed primarily at reproducing the metrical dimensions of the plant in a two- or three-dimensional graphic space. See the pioneering work of Hisao Honda (1971).

Growth unit Indicates the portion of stem that appears during a period of elongation. It may contain several internodes, since the latter are formed in the meristem and are only liable to elongation at a later point in time.

Internode Bud-less segment of plant stem between two nodes (the points of attachment of leaves on the stem).

Logical model Model based on the formal languages that were proposed in keeping with the logicist movement on reforming mathematics, such as that advanced by Alfred North Whitehead and Bertrand Russell in *Principia Mathematica*. In this regard, see the proposals of Joseph Henry Woodger (1937) and Aristid Lindenmayer (1964, 1968).

L-system Formal parallel rewriting grammar proposed by Aristid Lindenmayer in 1968. This sequential automaton was originally conceived in order to model cellular division. It is related to Chomsky's generative grammars, although it is not historically derived from them.

Mathematical model Formal model that typically takes the form of a functional equation or of a set of equations. By extension, it is a model in which one single set of axioms regulates the formalization of elements and their relationships. Example: Julian Huxley and Georges Teissier's (1936) law of allometry relating the relative growths of various dimensions of organs X and Y in a growing living being, in the form $Y = a \cdot X^b$. Allometry is a static mathematical model. A dynamic model includes time among its variables, in particular through the use of differential equations.

Maximal model According to agronomist and modeller Jean Bouchon (1995), a maximal or optimal model is a formal model that retains a large quantity of details, and is not restricted to just a very limited number of dimensions of the phenomenon under study. This is the case, according to Bouchon, of the universal models of the simulation of growing-plant architectures produced by AMAP.

Meristem   Embryonic cell tissue (located primarily in the growing tips of roots and shoots) that gives rise to primary or secondary growth.

Minimal model   According to agronomist and modeller Jean Bouchon (1995), a minimal model retains only what is strictly necessary to account for a particular evolution of the plant. It often retains a homogeneous mathematical form. The epistemology of minimal models, in this sense, characterized many schools of modelling in biology and ecophysiology for many years during the post-war period.

Monte-Carlo (method)   Calculation method suggested by Stanislaw Ulam in 1946 and put into practice by Ulam and Metropolis starting in 1949. This method consists of relying on draws of random (or pseudo-random) numbers in order to evaluate, by approximation (simulation of cases and then measurement of the results), certain deterministic or probabilistic functions that cannot be calculated analytically. The method is comparable with a technique of statistical sampling, but differs in that it can only be usefully processed by digital computer.

Morphogen   Term proposed by Alan Turing in 1952 to indicate the substances (whose existence was still hypothetical at that time) governing the development of forms in living beings, in particular because of their differing capacities for diffusion and reaction.

Morphogenetic field   Term proposed by a number of researchers in developmental biology and popularized by Julian S. Huxley (1932) in order to extend Charles Manning Child's (1915) notion of morphogenetic gradient. This rather polysemic term aims to indicate the causes (whether physical, chemical, positional or informational) of the succession of different organogenesis phenomena that are induced around a given initial group of cells during ontogenesis.

Multi-agent system (MAS)   A multi-agent system is a type of computerized formalization in which the "agents" – or in other words the computational entities defined in the programming language – share resources, communicate, interact and inherit their attributes over time. The emphasis is on the agents' autonomy. For further information see: J. Ferber: *Multi-Agent Systems: An Introduction to Distributed Artificial Intelligence*, Harlow: Addison Wesley Longman, 1999.

Multi-modelling   Type of pluriformalization that, by the 1970s, had been relatively fully standardized and developed by industrial-process engineers and modellers. This term (introduced by Tüncer Ibrahim Ören in 1989) indicates a simulation infrastructure that is capable of combining different types of modelling, including, at the outset, modelling with differential equations and discrete modelling. From 1991 it was taken up again and developed further, together with the notion of "multiformalism", in particular for application to other modelling objects.

Numerical simulation Numerical simulation differs from software-based simulation in that it remains a step-by-step and approximated algorithmic processing of a formal construct that possesses beforehand a mathematical unity and homogeneity of conception and construction. This prior unified mathematical formulation does not exist, however, either in the case of algorithmic simulations, which are sometimes awaiting a theory (cellular automata), or in the case of software-based simulations.

Object-oriented programming Type of programming that, from the end of the 1960s, and based on the SIMULA, SIMULA 67, C and C++ languages, broke with procedural programming in which the program was meant to be able to control from above the processes affecting the phenomena being modelled. By contrast, this type of programming defines the objects that then engage with other objects in accordance with the various attributes and behaviours that characterize them. In this way, the diversity of the objects that are modelled and the evolutive nature of their relationships can be even further taken into account. Object-oriented programming originated from the simulation of complex problems (e.g., the Minuteman missile) linking various heterogeneous elements. To a certain extent, it contrasts with the mathematical model approach, which was based entirely on a functional, procedural programming technique (procedure of calculation, of numerical analysis, of approximations, etc.) in which general languages aimed at mathematical formulations (such as FORTRAN) still sufficed. Object-oriented modelling opens the way to a potential *pluriformalization* of simulations, and tends to complete or replace monoformalized mathematicist conceptions. It *reifies* its concepts so as to be able to better replicate the phenomena it is simulating.

Phyllotaxis From the Greek φύλλον (*phùllon*), leaf, and τάξις (*tàxis*), order; indicating the relative arrangement of the leaves and branches of a plant and, by extension, any topological arrangement in a plant part (fruit, flower, etc.).

Pixel Contraction of *picture element*; smallest element or physical point in a digital image displayed on a bitmapped display.

Plant architecture An expression introduced in 1964 by botanists Francis Hallé and René Nozeran to indicate the structural morphological traits of plants, in contrast to their non-structural morphological traits.

Plant morphology Field of study in botany committed to describing both the external and the internal structures or shapes of plants.

Pluriformalization Generic term I have introduced to characterize all the formalization approaches that result in the coexistence of several formalisms and that are at the root of the recent complex computer simulations (such as multi-modelling and multi-agent systems). During computation, AMAP's first virtual plant simulations (1979–1987) typically involve and intertwine sub-models that are very different in type from a purely formal point of view: probabilistic, topological, logical (conditional branching) and geometrical.

Pragmatic model A model whose function is not essentially cognitive, but which is conceived with a view to the decisions to be taken and the operations to be carried out on a terrain that is generally considered complex. This is the case of the models implemented in France by DGRST starting in the 1970s, based on the work of biometrician Jean-Marie Legay and his colleagues.

Primary growth Increase in length of the stem.

Probabilistic model Model with a set of probabilistic axioms and rules, whose origins go back to the theory of errors in astronomy and to the theory of probable error of a mean on small sample sizes (Gosset, 1908), but whose systematic use in agronomy and biometry dates back to the introduction of the "hypothetical mathematical law" (called a "model" after the war) by Ronald A. Fisher in 1921. Postulating a probabilistic model thus appears to be necessary for the rigorous estimation of correlations between multivariable quantitative phenomena.

Process-based model Mathematical plant-growth model focused on formalization of the flows of energy and matter that are valid on an ecophysiological scale. These were one of the first mathematical models to be used in agronomy and silviculture. Their bird's-eye view mathematical approach has the drawback of assuming the homogeneity of the plant canopy (planted forests, monospecific forests, etc.)

Remathematization (phase) A term I introduced to describe the phase of maturity during which users of pluriformalized simulations feel the need and/or the necessity to remathematize (reunite, at least partially) the fragmented models and the programs that have been implemented to manage sub-models that, until that point, were mathematically heterogeneous.

Secondary growth Increase in thickness of the stem.

Shoot bending under self-weight loading Natural bending of some branches or shoots (in particular fruit-bearing ones) under the effect of weight, causing a compression that affects the flow of sap.

Software-based simulation A simulation is considered to be a "software-based simulation" when it relies on all the resources of computer programming, in particular an evolved language and object-oriented programming, in order to simultaneously process characteristics that differ in type, in time and in space. In such cases, the process cannot be represented in the form of a single step-by-step processing of a single algorithm (unless the aim is to consider only the level of machine language, although this has no simple and unequivocal translation, and is thus without any significance from a not-strictly-formal point of view, in the various expression languages specific to the sub-models being implemented and intertwined in the program in an evolved language). Even though this type of simulation has limitations, it is nonetheless particularly adapted to the simulation of composite objects. It should be distinguished from algorithmic simulation.

Structural-functional model Recent type of plant model (late 1990s) that combines the process-based approach (see process-based model) and the architectural-model simulation approach (see universal simulation) so as to take account of the feedback effect of physiological functioning on the plant's geometry and topology.

Sub-symbolic It may be said that computer simulations make a sub-symbolic use of the symbols that they process using the computer, insofar as these simulations are not based primarily on the formal symbols' ability to give rise to conventional rules of condensation, combination and transformation. On the contrary, these rules may be broken down in the simulation and once again become highly iconic, as in the discretized neutron-by-neutron simulation of a nuclear charge. This does not mean, however, that the simulation replaces the symbolic processing by anything else, but rather that the symbols that it uses are not entirely processed and used as such. Thus, we remain in the realm of the systems of symbols, and it is often one system of symbols that simulates the behaviour of another system of symbols. This may often be, for example, a (discrete) set of axioms that simulates another (continuous) set of axioms, with one set of axioms thus being considered sub-symbolic *compared* with the other. For further information in this regard, see Phan and Varenne, 2010.

Supra-simulation A term I introduced in order to qualify a virtual experiment carried out on the simulation of a scene replication that is, in its own right, already complete, calibrated and stabilized. For example, in the work of Jean Dauzat (Dauzat and Hautecoeur, 1991), the simulation of reflectance is carried out by simulating virtual solar ray-tracings on a virtual forest that is itself the result of a series of architectural simulations of single trees.

Theory In the context of plant modelling, the theory contrasts with the model insofar as it ranks on the level of principles (in particular in Rashevsky) because, according to some, it retains a reflection of the very essence of the phenomena and especially because it may give rise to a certain number of deduced results in the form of theorems (see Françon with regard to the usefulness of combinatorics in this regard). Many of the formalisms used later as simple models were introduced at first as the parts of some ambitious essentialist theory: this was the case of the L-systems that, in 1968, seemed to some extent to bring parts of developmental biology into the ranks of the theorematic disciplines.

Universal simulation An expression I use to qualify a type of software-based simulation that offers a possibility of configuration that makes it capable of simulating the entirety of the potential manifestations of a given object or system. For example, de Reffye's architectural simulation of plants (1979) is universal to the extent that it is not specific to one single architectural model of plants: in fact, it is capable of simulating all of the 24 architectural models that are known to exist on Earth.

Validation For a formal model, validation indicates the set of the procedures (comparison with measurements, coherence tests, statistical tests) that are implemented in order to ensure that the model is admissible, taking into account the epistemic function that is intended to be assigned to it (use for experimentation, representation, explanation, comprehension, integration, action, decision-making or consultation: see Chapter 8).

Variance analysis See "variance reduction".

Variance reduction The technique of variance reduction or variance analysis is based on the principle that the variances *combine* when the effects of the different processing or factors that act on a plant (or on any other multivariate phenomenon) are independent. The effects of the variables must thus be additive. In this case, it can be seen that it is possible to analyse the variance in the sense that the inter-category variance in particular can be distinguished from the intra-category or residual variance.

Verification   For a formal model, the set of test and control procedures that make it possible to evaluate the quality (harmlessness) of the computer implementation of the model with respect to the performance expected from that model.

Virtual experiment or simulation experiment   Type of experiment carried out by computer simulation so as to ensure that the results it produces have an epistemic status that can be considered comparable to those obtained from actual experiments (they may, in particular, be used to disprove a theoretical model: Wagensberg, 1985). This expression is well documented in the literature on simulation in the engineering and natural sciences, in particular. The term "virtual laboratory", however, occurs primarily in the case of forecasting. The reality and the source of the "empirical" nature of these experiments are still debated (see Conclusion), especially when the simulation is no longer simply an approximate calculation of a unique mathematical model.

Virtual laboratory   Software environment that allows various virtual experiments of either scientific or simply pedagogical aspiration to be conducted. See the work of Przemysław Prusinkiewicz (1989–1990). A virtual laboratory differs from just virtual experimentation in that it claims to be a complete environment that is valid for a wide range of virtual experiments.

Voxel   The equivalent, for volumes, of the pixel. This is the smallest unit of volume visible on screen.

Water efficiency law   This phenomenological law, known to ecophysiologists since the 1930s, postulates that the matter created during photosynthesis is continuously proportional to the plant's transpiration.

Selected bibliography

Abelson (H.), "Logo graphics as a mathematical environment", *Proceedings of the Annual Conference of the Association of Computing Machinery, ACM-CSC-ER*, Houston, ACM-Press, 1976.

Aguilar-Marin (J.) (Ed.), "Modélisation mathématique et simulation des systèmes de l'environnement", *Papers of 1st Seminar of the Programme Interdisciplinaire de Recherche sur l'Environnement du CNRS* [CNRS Interdisciplinary Research Programme on the Environment], Toulouse, Éditions du CNRS, 1982.

Althusser (L.), *Philosophie et philosophie spontanée des savants*, Paris, Maspero, 1974.

Amblard (F.), "Comprendre le fonctionnement de simulations sociales individus-centrées: application à des modèles de dynamiques d'opinion", Computer science thesis, Université Blaise Pascal – Clermont II, 2003.

Amblard (F.) and Phan (D.), *Modélisation et simulation multi-agents: applications aux sciences de l'homme et de la société*, Amblard, F. and Phan D. (Eds), Paris, Hermes, 2006.

Andrieu (B.) (Ed.), "Modélisation architecturale du fonctionnement des cultures: Orientations de l'équipe de Bioclimatologie de Grignon", in *Modélisation architecturale*, B. Andrieu (Ed.), proceedings of seminar of 10–12 March 1997, Bioclimatology Department, INRA-Grignon, 15–18.

Aono (M.), Kunii (T.L.), "Botanical tree image generation", *IEEE Computer Graphics and Applications*, 1984, Vol. 4, No. 5, May, 10–34.

Apter (M.J.), *Cybernetics and Development*, Oxford, Pergamon Press, 1966.

Armatte (M.), Dahan-Dalmedico (A.) (Eds), "Modèles et modélisations 1950–2000", special edition of *Revue d'histoire des sciences [History of science journal]*, 2004, Vol. 57, No. 2, July–December.

Badiou (A.), *Le concept de modèle*, Paris, Maspero, 1969.

Barker (S.B.), Cumming (G.), Horsfield (K.), "Quantitative morphometry of the branching structures of trees", *Journal of Theoretical Biology*, 1973, Vol. 40, 33–43.

Barthélémy (D.), "Levels of organization and repetition phenomena", *Acta Biotheoretica*, 1991, Vol. 39, Nos 3–4, 309–323.

Barthélémy (D.), "Architecture et sexualité chez quelques plantes tropicales: le concept de floraison automatique", Doctoral thesis, Physiology, Biology of Organisms and Population Biology, University of Montpellier, 1988.

Barthélémy (D.), Blaise (F.), Fourcaud (T.), Nicolini (E.), "Modélisation et simulation de l'architecture des arbres: bilan et perspectives", *Revue Forestière Française* [French Forestry Journal], special edition, 1995, Vol. 47, 71–96.

Barthélémy (D.), Edelin (C.), Hallé (F.), "Architectural concepts for tropical trees", in L.B. Holm-Nielsen, I. Nielsen, H. Balslev (Eds), *Tropical Forests: Botanical Dynamics, Speciation and Diversity*, London, Academic Press, 1989, 89–100.

Batty (M.), *Cities and Complexity*, Cambridge, The MIT Press, 2005.

Bauer (P.S.), "The validity of minimal principles in physiology", *The Journal of General Physiology*, 1930, Vol. 13, July, 617–619.

Beaumont (J.H.), "An analysis of growth and yield relationships of coffee trees in the Kona district, Hawaii", *Journal of Agricultural Research (Washington)*, 1939, Vol. 59, No. 3, 1 August, 223–235.

Bedau (M.A.), "Philosophical content and method of artificial life", in T.W. Binum, J.H. Moor, *The Digital Phoenix: How Computers are Changing Philosophy*, Oxford, Basil Blackwell, 1998, 135–152.

Bell (A.D.), "Computerized vegetative mobility in rhizomatous plants", in A. Lindenmayer, G. Rozenberg (Eds), *Automata, Languages, Development*, Amsterdam, North-Holland Publishing Company, 1976, 3–14.

Black (M.), *Models and Metaphors: Studies in Language and Philosophy*, Ithaca and London, Cornell University Press, 1962; sixth printing: 1976.

Blaise (F.), "Simulation du parallélisme dans la croissance des plantes et application", new thesis No. 1071 (computer science specialization), Strasbourg, Université Louis Pasteur, 1991.

Blaise (F.), "Simulation de couverts végétaux réalistes en 3D", in D.C./L.W. Fritz, J.R. Lucas (Eds), *XVIIth I.S.P.R.S. Congress*, Washington, International Archives of Photogrammetry and Remote Sensing, 1992, 24, B3, (3), 207–212.

Blasco (F.) (Ed.), "Tendances nouvelles en modélisation pour l'environnement", *Proceedings of the Symposium of 16–17 January 1996: Programme Environnement du CNRS*, Paris, éditions Elsevier, 1997.

Blaise (F.), Barczi (J.F.), Jaeger (M.), Dinouard (P.), Reffye (de) (P.), "Simulation of the growth of plants: modeling of metamorphosis and spatial interactions in the architecture and development of plants", in T.L. Kunii, L. Luciani (Eds) *Cyberworlds*, Tokyo, Springer Verlag, 1998, 81–109.

Blaise (F.), Reffye (de) (P.), "Simulation de la croissance des arbres et influence du milieu: le logiciel AMAPpara", in J. Tankoano (Ed.), *Proceedings of the 2nd African Conference on Research, Computer Science (CARI '94)*, Ouagadougou, Burkina Faso, 12–18 October 1994, INRIA-ORSTOM, 1994, 61–75.

Bouchon (J.) (Ed.), *Architecture des arbres fruitiers et forestiers*, Paris, INRA-éditions, "Les colloques" collection, No. 74, 1995.

Bouchon (J.), Poupardin (D.), "Entretien avec Jean Bouchon", 21 July 1995, *Nancy, with Denis Poupardin on behalf of Archorales-INRA (Archives orales de l'INRA [INRA oral archives])*, 21 pages.

Bouchon (J.), Reffye (de) (P.), Barthélémy (D.), *Modélisation et simulation de l'architecture des végétaux*, Paris, INRA-éditions, "Science Update" collection, 1997.

Bouleau (N.), *Philosophies des mathématiques et de la modélisation*, Paris, l'Harmattan, 1999.

Bouligand (Y.) (Ed.), *La morphogenèse: de la biologie aux mathématiques*, Paris, Maloine, 1980.

Boumans (M.), "Built-in justification", in M. S. Morgan, M. Morrison (Eds), *Models as Mediators*, Cambridge, Cambridge University Press, 1999, 66–96.

Boutot (A.), *L'invention des formes*, Paris, Odile Jacob, 1993.

Brissaud (M.), Forsé (M.), Sighed (A.) (Eds), "La modélisation, confluent des sciences", *Proceedings of the Interdisciplinary Symposium of 15 and 16 June 1989 at Villeurbanne*, Paris, Éditions du CNRS, 1990.

Buck-Sorlin (G.H.), Bachmann (K.), "Simulating the morphology of barley spike phenotypes using genotype information", *Agronomie*, 2000, Vol. 20, 691–702.

Burks (A.W.) (Ed.), *Essays on Cellular Automata*, Urbana, University of Illinois Press, 1970.

Canguilhem (G.), "Modèles et analogies dans la découverte en biologie", *Études d'histoire et de philosophie des sciences concernant les vivants et la vie*, Paris, Vrin, 1968; reprint: 1994, 305–318.

Canguilhem (G.), *Idéologie et rationalité*, Paris, Vrin, 1977; reprint: 2000.

Capot (J.), "L'amélioration du caféier en Côte d'Ivoire: Les hybrides Arabusta", *Café, Cacao, Thé*, 1972, Vol. 16, No. 1, January–March, 3–16.

Carnap (R.), "Die physikalische Sprache als Universalsprache der Wissenschaft", *Erkenntnis*, Bd. 2, H 5/6, 1932, 432–465.

Carnap (R.), *Logische Syntax des Sprache,* Vienna, Julius Springer, 1934; *The Logical Syntax of Language*, trans. A.S. von Zeppelin, London, Paul Kegan, 1937; reprint: London, Open Court, 2002.

Cartwright (N.), *A Dappled World: A Study of the Boundaries of Science*, Cambridge, Cambridge University Press, 1999.

Carusi (A.), Hoel (A.S.), Webmoor (T.), Woolgar (S.), *Visualization in the Age of Computerization*, New York, Routledge, 2015.

Caseau (P.), "Les modèles numériques et leur place dans la recherche-développement", *Culture Technique*, 1988, No. 18, March, 126–130.

Chaunu (P.), *Le temps des réformes*, Paris, Fayard, 1975.

Chazal (G.), "La pensée et les machines: le mécanisme algorithmique de John von Neumann", *preface to Théorie générale et logique des automates*, J. von Neumann, Champ-Vallon, coll. Milieux, 1996, 7–58.

Chazal (G.), "La simulation informatique comme mesure du possible", in Centre d'Analyse des Formes et des Systèmes de la Faculté de Philosophie de Lyon III (Ed.), *La mesure: instruments et philosophes*, Paris, Champ-Vallon, 1994, 147–155.

Chorafas (D.N.), *La simulation mathématique et ses applications*, Paris, Dunod, 1966.

CIRAD-1997, *Le Cirad en 1997,* company review, Service des éditions du CIRAD, Délégation à l'information scientifique et technique, Svi-Publicep, Montpellier, 1997.

CIRAD-1998, *Images de la recherche,* Service des éditions du CIRAD, Délégation à l'information scientifique et technique, Svi-Publicep, Montpellier, 1999.

CIRAD-1999, *Le Cirad en 1998,* company review, Service des éditions du CIRAD, Délégation à l'information scientifique et technique, Svi-Publicep, Montpellier, 1999.

Cohen (D.), "Computer simulation of biological pattern generation processes", *Nature*, 1967, Vol. 216, 21 October, 246–248.

Cooper (N.G.) (Ed.), *From Cardinals to Chaos,* Los Alamos Science, special issue, 1987; reprint: Cambridge, Cambridge University Press, 1989.

Coquillard (P.), Hill (D.R.C.), *Modélisations et simulations d'écosystèmes*, Paris, Masson, 1997.

Corner (E.J.H.), *The Life of Plants*, Cleveland, World Publisher, 1964, trans. Paule Corsin: *La vie des* plantes, Paris, Stock, 1970.

Costes (E.), Reffye (de) (P.), Lichou (J.), Guédon (Y.), Audubert (A.), Jay (M.), "Stochastic modelling of apricot growth units and branching", *3rd International Symposium on Computer Modelling in Fruit Research and Orchard Management*, New Zealand, Palmerston North, *Acta Horticulturae*, 1992, Vol. 313, 89–98.

Cournède (P.H.), "Dynamic system of plant growth", Habilitation Doctorate, University of Montpellier, 2009, https://tel.archives-ouvertes.fr/tel-00377462v2.

Cournède (P.H.), Kang (M.Z.), Mathieu (A.), Barczi (J.F.), Yan (H.P.), Hu (B.G.), Reffye (de) (P.), "Structural factorization of plants to compute their functional and architectural growth", *Simulation*, 2006, Vol. 82, No. 7, 427–438.

Cox (D.R.), *Planning of Experiments*, New York, Wiley and Sons, 1958.

Cox (D.R.), *Renewal Theory*, London, Chapman and Hall, UK, 1962.

Cox (D.R.), Lewis (P.A.W.), *The Statistical Analysis of Series of Events*, London, Methuen and Co. Ltd, 1966, trans.: *L'analyse statistique des séries d'événements*, Paris, Dunod, 1969.

Dahan-Dalmedico (A.), "Modèles et modélisations: le foisonnement des pratiques contemporaines exige une réflexion théorique nouvelle", *Letter from SPM (Département des Sciences Physiques et Mathématiques* [Physical Sciences and Mathematics Department], CNRS), No. 42, December 2003, 26–28.

Damien (R.), *Bibliothèque et État*, Paris, PUF, 1995.

Dauzat (J.), "Radiative transfer simulation on computer models of *Elaeis guineensis*", *Oléagineux*, 1994, Vol. 49, No. 3, 8–90.

Dauzat (J.), Hautecoeur (O.), "Simulation des transferts radiatifs sur maquettes informatiques de couverts végétaux", in J.J. Hunt (Ed.), *Physical Measurements and Signatures in Remote Sensing*, European Space Agency (ESA), Courchevel, France, 1991, 415–418.

Delattre (P.), Thellier (M.), *Élaboration et justification des modèles*, Paris, Maloine, 1979, 2 volumes.

Deléage (J.P.), *Histoire de l'écologie*, Paris, La Découverte, 1991.

Deleuze (Ch.), "Pour une dendrométrie fonctionnelle: essai sur l'intégration des connaissances écophysiologiques dans les modèles de production ligneuse", thesis for Université Claude Bernard, Lyon I.

Desrosières (A.), *La politique des grandes nombres. Histoire de la raison statistique*, Paris, La Découverte & Syros, 1993.

Di Paolo (E.A.) *et al.*, "Simulation models as opaque thought experiments", in M.A. Bedau *et al.* (Eds) *Artificial Life VII*, Proceedings of the 7th International Conference on Artificial Life, Cambridge, MIT Press, 2000, 497–506.

Dietrich (M.R.), "Monte-Carlo experiments and the defense of diffusion models in molecular population genetics", *Biology and Philosophy*, 1996, Vol. 11, No. 3, July, 339–356.

Douady (S.), Couder (Y.), "Phyllotaxis as a physical self-organized growth process", *Physical Review Letters*, 1992, Vol. 68, No. 13, 2098–2101.

Duboz (R.), "Intégration de modèles hétérogènes pour la modélisation et la simulation de systèmes complexes – Application à la modélisation multi-échelles en écologie marine", computer science thesis, Université du Littoral, 2004.

Dubucs (J.), 2002, "Simulations et modélisations", *Pour la science*, No. 300, October 2002.

Duhem (P.), *La théorie physique – son objet – sa structure*, Paris, 1906–1914; reprint: Vrin, 1989.

Duhem (P.), *Sauver les phénomènes*, Paris, 1908; reprint: Vrin, 1994.

Dupré (J.), *The Disorder of Things: Metaphysical Foundations of the Disunity of Science*, Cambridge, Harvard University Press, 1993.

Dzierzon (H.), Kurth (W.), "LIGNUM: a Finnish tree growth model and its interface to the French AMAPmod database", in F. Hölker (Ed.), *Scales, Hierarchies and Emergent Properties in Ecological Models*, Frankfurt am Main, Peter Lang, 2002, 29–46.

Edelin (C.), *Images de l'architecture des conifères*. 3rd cycle Doctoral thesis in plant biology, University of Montpellier II, 1977.

Eden (M.), "A probabilistic model for morphogenesis", *Symposium on Information Theory in Biology*, New York, Pergamon Press, 1958, 359–370.

Eden (M.), "A two-dimensional growth process", *Fourth Berkeley Symposium on Mathematical Statistics and Probability*, Berkeley, University of California Press, 1960, 223–239.

Emmeche (C.), *The Garden in the Machine*, Princeton, Princeton University Press, 1994.

Erickson (R.O.), "Relative elemental rates and anisotropy of growth in area: a computer programme", *Journal of Experimental Botany*, 1966, Vol. 17, No. 51, May, 390–403.

Erickson (R.O.), "Modeling of plant growth", *Annual Review of Plant Physiology*, 1976, Vol. 27, 407–434.

Feltz (B.), *Croisées biologiques*, Brussels, éditions Ciaco, 1991.

Feltz (B.), Crommelinck (M.), Goujon (P.) (Eds), *Auto-organisation et émergence dans les sciences de la vie*, Brussels, Ousia, 1999.

Ferber (J.), *Les systèmes multi-agents: vers une intelligence collective*, Paris, InterEditions, 1995.

Ferber (J.), *Multi-Agent Systems: An Introduction to Distributed Artificial Intelligence*, Harlow, Addison Wesley Longman, 1999.

Feuvrier (C.V.), *La simulation des systèmes*, Paris, Dunod, 1971.

Fisher (J.B.), "How predictive are computer simulations of tree architecture?", *International Journal of Plant Sciences*, 1992, Vol. 153, No. 3, 137–146.

Fisher (J.B.), Honda (H.), "Computer simulation of branching pattern and geometry in *Terminalia (Combretaceae)*, a tropical tree", *Botanical Gazette*, 1977, Vol. 138, No. 4, 377–384.

Fisher (J.B.), Honda (H.), "Branch geometry and effective leaf area: a study of *Terminalia* branching pattern, 1. theoretical trees", *American Journal of Botany*, 1979, Vol. 66, No. 6, 633–644.

Fisher (J.B.), Honda (H.), "Branch geometry and effective leaf area: a study of *Terminalia* branching pattern, 2. survey of real trees", *American Journal of Botany*, 1979, Vol. 66, No. 6, 645–655.

Fisher (R.A.), "Some remarks of the methods formulated in a recent article on 'the quantitative analysis of plant growth'", *Annals of Applied Biology*, 1921, Vol. 7, 367–372.

Fisher (R.A.), "Studies in crop variation. I. An examination of the yield of dressed grain from Broadbalk", *Journal of Agricultural Sciences*, 1921, Vol. 11, No. 2, 107–135.

Fisher (R.A.), "On the mathematical foundations of theoretical statistics", *Philosophical Transactions of the Royal Society of London*, A, 1922, Vol. 222, 309–368.

Fisher (R.A.), *Statistical Methods for Research Works*, Edinburgh, Oliver and Boyd, 1925.

Fleury (V.), Gouyet (J.F.), Léonetti (M.), *Branching in Nature*, Berlin, Springer, 2001.

Forget (A.), "La modélisation", in B. Gauthier, I. Bourgeois (Eds), *Recherche sociale: de la problématique à la collecte des données* [Social research: from problems to data collection], 6th edition, Québec, Presses de l'Université du Québec, 2016, Chapter 6, 129–158.

Fournier (C.), "Modélisation des interactions entre plantes au sein des peuplements. Application à la simulation des régulations de la mophogenèse aérienne du maïs (*Zea mays L.*) par la compétition pour la lumière", thesis, Institut National Agronomique Paris-Grignon, April 2000.

Fournier (C.), Andrieu (B.), "Utilisation de l'approche L-système pour la modélisation architecturale du développement du maïs", in B. Andrieu (Ed.), *Modélisation architecturale [Architectural modelling]*, Proceedings of the seminar of 10–12 March 1997, Bioclimatology Department, INRA-Grignon, 203–211.

Fournier (C.), Andrieu (B.), "A 3D architectural and process-based model of maize development", *Annals of Botany*, 1998, Vol. 81, No. 2, 233–250.

Fournier (C.), Andrieu (B.), "Dynamics of the elongation of the internodes in maize (*Zea mays L.*): analysis of phases of elongation and their relationships to phytomer development", *Annals of Botany*, 2000, Vol. 86, No. 3, 551–563.

Franc (A.), Gourlet-Fleury (S.), Picard (N.), *Une introduction à la modélisation des forêts hétérogènes*, Nancy, ENGREF éditions, 2000.

Françon (J.), "Arbres et nombres de Strahler dans diverses sciences", *Revue du Palais de la Découverte [Palais de la Découverte journal]*, 1984, Vol. 12, No. 120, 29–36.

Françon (J.), "Sur la modélisation informatique de l'architecture et du développement des végétaux", in C. Edelin (Ed.), *Naturalia Monspeliensa*, Special Edition, Montpellier, 1991, A7, 231–247.

Françon (J.), Lienhardt (P.), "Basic principles of topology-based methods for simulating metamorphosis of natural objects", in N.M. Thalmann, D. Thalmann (Eds), *Artificial Life and Virtual Reality*, Chichester, John Wiley and Sons, 1994, 23–44.

Freudenthal (H.) (Ed.), *The Concept and the Role of the Model in Mathematics and Natural and Social Sciences*, Dordrecht, D. Reidel Publishing Company, 1961.

Galison (P.), *How Experiments End*, Chicago, Chicago University Press, 1987.

Galison (P.), *Image and Logic*, Chicago, University of Chicago Press, 1997.

Galison (P.), Stump (D.J.) (Eds), *The Disunity of Science*, Stanford, Stanford University Press, 1996.

Garfinkel (D.), "Digital computer simulation of ecological systems", *Nature*, 1962, Vol. 194, June, 856–857.

Gayon, (J.), "History of the concept of allometry", *American Zoologist*, 2000, Vol. 40, No. 5, 748–758.

Gigerenzer (G.), Swijtnik (Z.), Porter (T.), Daston (L.), Beatty (J.), Krüger (L.) (Eds), *The Empire of Chance: How Probability Changed Science and Everyday Life*, Cambridge, Cambridge University Press, Ideas in Context series, 1989; reprint: 1997.

Gilbert (S.F.), "Cellular politics: Ernest Everett Just, Richard B. Goldschmidt, and the attempt to reconcile embryology and genetics", in R. Rainger, K.R. Benson, J. Maienschein (Eds), *The American Development of Biology*, Philadelphia, University of Pennsylvania Press, 1988, 311–346.

Godin (C.), Caraglio (Y.), "A multiscale model of plant topological structures", *Journal of Theoretical Biology*, 1998, Vol. 191, No. 1, 1–46.

Godin (C.), Guédon (Y.), Costes (E.), "Exploration of a plant architecture database with the AMAPmod software illustrated on an apple tree hybrid family", *Agronomie*, 1999, Vol. 19, No. 3/4, 163–184.

Goethe (von) (J.W.), *La métamorphose des plantes, 1790–1807*, trans. H. Bideau, 1975, Paris, éditions Triades; republished: 1992.

Goodman (N.), *Languages of Art: An Approach to a Theory of Symbols*, Indianapolis, Bobbs-Merrill, 1976; 1st edition: 1968.

Goodman (N.), "Routes of reference", *Critical Inquiry*, 1981, Vol. 8, No. 1, 121–132.

Gorenflot (R.), *Biologie végétale: plantes supérieurs: appareil végétatif*, Paris, Masson; 1st edition: 1977; 2nd edition: 1998.

Gosset (alias "Student") (W.S.), "The probable error of a mean", *Biometrika*, 1908, Vol. 6, No. 1, 1–25.

Goujon (P.), "La biologie à l'ère de l'informatique. Connaissance et naissance de la vie artificielle, première partie", *Revue des Questions Scientifiques [Journal of Scientific Matters]*, 1994, Vol. 165, No. 1, 53–84.

Goujon (P.), "La biologie à l'ère de l'informatique. Connaissance et naissance de la vie artificielle, seconde partie", *Revue des Questions Scientifiques [Journal of Scientific Matters]*, 1994, Vol. 165, No. 2, 119–153.

Gramelsberger (G.) (Ed.), *From science to computational sciences*, Zürich, Diaphanes, 2011.

Granger (G.G.), *Le probable, le possible et le virtuel*, Paris, Odile Jacob, 1995.

Granger (G.G.), "Simuler et comprendre", *Philosophie, langage, science*, Paris, EDP, 2003, 187–192.

Greene (N.), "Voxel space automata: modelling with stochastic growth processes in voxel space", *Computer Graphics*, 1989, Vol. 23, No. 3, 175–184.

Gregg (J.R.), Harris (F.T.C.) (Eds), *Form and Strategy in Science: Studies Dedicated to Joseph Henry Woodger on the Occasion of his Seventieth Birthday*, Dordrecht, D. Reidel Publishing Company, 1964.

Grimm (V.), "Ten years of individual-based modeling in ecology: what have we learned and what could we learn in the future?", *Ecological Modelling*, 1999, Vol. 115, Nos 2–3, 129–148.

Gruntman (M.), Novoplanksy (A.), "Ontogenetic contingency of tolerance mechanisms in response to apical damage", *Annals of Botany*, 2011, Vol. 108, No. 5, 965–973.

Guédès (M.), "La théorie de la métamorphose en morphologie végétale: des origines à Goethe et Batsch", *Revue d'histoire des sciences appliquées* [History of applied sciences journal], 1969, Vol. 22, No. 4, 323–363.

Guédès (M.), "La théorie de la métamorphose en morphologie végétale: A.P. de Candolle et P.J.F. Turpin", *Revue d'histoire des sciences appliquées* [History of applied sciences journal], 1972, Vol. 25, No. 3, 253–270.

Guédès (M.), "La théorie de la métamorphose en morphologie végétale. La métamorphose et l'idée d'évolution chez Alexandre Braun", *Epistémè*, 1973, Vol. 7, 32–51.

Guillevic (P.), "Modélisation des bilans radiatif et énergétique des couverts végétaux", specialization "télédétection de la biosphère continentale – modélisation" [remote sensing of the continental biosphere – modelling], Doctoral thesis at Université Paul Sabatier, December 1999.

Hacking (I.), "Do we see through a microscope?", *Pacific Philosophical Quarterly*, 1981, Vol. 62, No. 4, 305–322.

Hacking (I.), *Representing and Intervening*, Cambridge, Cambridge University Press, 1983; trans.: *Concevoir et expérimenter*, Paris, Bourgois, 1989.

Hacking (I.), *The Taming of Chance*, Cambridge, Cambridge University Press, 1990.

Hallé (F.), "Modèles architecturaux chez les arbres tropicaux", in Delattre (P.), Thellier (M.), *Élaboration et justification des modèles*, Vol. 2, Paris, Maloine, 1979, 537–550.

Hallé (F.), *Éloge de la plante. Pour une nouvelle biologie*, Paris, Seuil, 1999.

Hallé (F.), Oldeman (R.A.A.), *Essai sur l'architecture et la dynamique de croissance des arbres tropicaux*, Paris, Masson, 1970.

Hallé (F.), Oldeman (R.A.A.), Tomlinson (P.B.), *Tropical Trees and Forests. An Architectural Analysis*, New York, Springer Verlag, 1978.

Harris (T.E.), *The Theory of Branching Processes*, Vol. 119 of the "Die Grundlehren der mathematischen Wissenschaften" series, Berlin, Springer-Verlag, 1963.

Hartmann (S.), "Simulation", *Enzyklopädie Philosophie und Wissenschaftstheorie*, Vol. 3, Stuttgart, Verlag Metzler, 1995, 807–809.

Hartmann (S.), "The world as a process: simulation in the natural and social sciences", in R. Hegselmann, U. Muller, K. Troitzsch (Eds), *Modelling and Simulation in the Social Sciences from the Philosophy of Science Point of View*, Dordrecht, Kluwer Academic, 1996, 77–100.

Hesse (M.B.), *Models and Analogies in Science*, Notre Dame, University of Notre Dame Press, 1966; 2nd printing: 1970.

Heudin (J.C.), *La vie artificielle*, Paris, Hermès, 1994.

Hill (D.R.C.), "Verification and validation of ecosystem simulation models", in *Proceedings of the Summer Simulation Conference*, 24–26 July, Ottawa, 1995, 176–182.

Hill (D.R.C.), *Object Oriented Analysis and Simulation*, Boston, Addison Wesley, 1996.

Hill (D.R.C.), "Contribution à la modélisation de systèmes complexes: application à la simulation d'écosystèmes", mémoire d'habilitation à diriger des recherches [authorization to conduct research thesis] for Université Blaise Pascal, computer science specialization, Clermont-Ferrand, 2000.

Hodges (A.), *Alan Turing: The Enigma of Intelligence*, London, Burnett Books Ltd in association with the Hutchinson Publishing Group, 1983; abridged translation: *Alan Turing ou l'énigme de l'intelligence*, Paris, Payot, 1988.

Honda (H.), "Description of the form of trees by the parameters of the tree-like body: effects of the branching angle and the branch length on the shape of the tree-like body", *Journal of Theoretical Biology*, 1971, Vol. 31, No. 2, 331–338.

Honda (H.), "Pattern formation of the coenobial algae Pediastrum biwae Negoro", *Journal of Theoretical Biology*, 1973, Vol. 42, No. 3, 461–481.

Honda (H.), Eguchi (G.), "How much does the cell boundary contract in a monolayered cell sheet?", *Journal of Theoretical Biology*, 1980, Vol. 84, No. 3, 575–588.

Honda (H.), Fisher (J.B.), "Tree branch angle: maximizing effective leaf area", *Science*, 1978, Vol. 199, No. 4331, 24 February, 888–890.

Honda (H.), Fisher (J.B.), "Ratio of tree branch lengths: the equitable distribution of leaf clusters on branches", *Proceedings of the National Academy of Sciences of the USA: Botany*, 1979, Vol. 76, No. 8, August, 3875–3879.

Honda (H.), Hatta (H.), Fisher (J.B.), "Branch geometry in *Cornus Kousa* (Cornaceae): computer simulations", *American Journal of Botany*, 1997, Vol. 84, No. 6, 745–755.

Honda (H.), Tomlinson (P.B.), Fisher (J.B.), "Computer simulation of branch interaction and regulation by unequal flow rates in botanical trees", *American Journal of Botany*, 1981, Vol. 68, No. 4, 569–585.

Honda (H.), Tomlinson (P.B.), Fisher (J.B.), "Two geometrical models of branching of botanical trees", *Annals of Botany*, 1982, Vol. 49, No. 1, 1–11.

Houllier (F.), "Modélisation de la dynamique des peuplements forestiers. Relations entre objectifs, structures, données et méthodes", in J. Demongeot, P. Malgrange (Eds), *Biologie et Économie*, Dijon, Librairie de l'Université de Bourgogne, 1987, 271–293.

Houllier (F.), "Dynamique des peuplements de forêt dense humide: dialogue entre éco-logues, expérimentateurs et modélisateurs", *Revue d'Écologie*, 1995, Vol. 50, No. 3, 303–311.

Houllier (F.), Bouchon (J.), Birot (Y.), "Modélisation de la dynamique des peuplements forestiers: état et perspectives", *Revue Forestière Française* [French Forestry Journal], 1991, Vol. 43, No. 2, 87–108.

Houllier (F.), Leban (J.M.), Colin (F.), "Linking growth modelling to timber quality assessment for Norway spruce", *Forest Ecology and Management*, 1995, Vol. 74, Nos 1–3, 91–102.

Humphreys (P.), "Computer simulations", *PSA (Philosophy of Science Association)*, 1990, Vol. 2, 497–506.

Humphreys (P.), "Computational models", *Philosophy of Science*, 2002, Vol. 69, September, S1–S11.

Humphreys (P.), *Extending Ourselves: Computational Science, Empiricism and Scientific Method*, Oxford, Oxford University Press, 2004.

Humphreys (P.), "The philosophical novelty of computer simulation methods", *Synthese*, 2009, Vol. 169, No. 3, 615–626.

Israel (G.), *La mathématisation du réel*, Paris, Seuil, 1996.

Jaeger (M.), "Représentation et simulation de croissance des végétaux", new thesis No. 1071 (specialization in computer science), Strasbourg, Université Louis Pasteur, 1987.

Jaeger (M.), Reffye (de) (P.), "Basic concepts of computer simulation of plant growth", *The Journal of Biosciences*, 1992, Vol. 17, No. 3, 275–291.

Jakobson (R.), Halle (M.), "Phonology and phonetics, première partie" in *Fundamentals of Language*, The Hague, Mouton & Co., 1956; modified and translated: "Phonologie et phonétique", *Essais de linguistique générale*, Paris, Éditions de Minuit, 1963, 103–149.

Jean (R.V.), *Phytomathématique*, Montreal, Les Presses de l'Université du Québec, 1978.

Jean (R.V.), *Phyllotaxis – A Systemic Study in Plant Morphogenesis*, Cambridge, Cambridge University Press, 1994; reprint: 1995.

Kang (M.Z.), Reffye (de) (P.), Barczi (J.F.), Hu (B.G.), "Fast algorithm for stochastic tree computation", 11th International Conference in Central Europe on Computer Graphics: Visualization and Computer Vision 2003, in cooperation with *EUROGRAPHICS, Journal of WSCG (Winter School of Computer Graphics)*, 2003, Vol. 11, No. 1, 8.

Kant (E.), *Critique de la Faculté de Juger*, 1790, trans. J.H. Bernard, London, Macmillan and Co. Ltd, 1914.

Kauffman (S.), *At Home in the Universe. The Search for the Laws of Self-Organization and Complexity*, Oxford, Oxford University Press, 1995.

Keller (E.F.), *Refiguring Life. Metaphors of Twentieth-Century Biology*, New York, Columbia University Press, 1995.

Keller (E.F.), "Models, simulation and 'computer experiments'", in Hans Radder (Ed.), *The Philosophy of Scientific Experimentation*, Pittsburgh, University of Pittsburgh Press, 2002, 198–215.

Keller (E.F.), *Making Sense of Life. Explaining Biological Development with Models, Metaphors and Machines*, Cambridge, Harvard University Press, 2002; 2nd edition: 2003.

Kingsland (S.E.), *Modeling Nature: Episodes in the History of Population Ecology*, Chicago and London, The University of Chicago Press, 1985; second edition with a new postface: same publisher, 1995.

Kostitzin (V.A.), *Biologie mathématique*, Paris, Armand Colin, 1937.

Kurth (W.), "Morphological models of plant growth: possibilities and ecological relevance", *Ecological Modelling*, 1994, Vols 75/76, 299–308.

Kurth (W.), "Stochastic sensitive growth grammars: a basis for morphological models of tree growth", *Proceedings of symposium on L'arbre: Biologie et Développement [The tree: Biology and Development]*, University of Montpellier, 11–16 September 1995.

Kurth (W.), "Spatial structure, sensitivity and communication in rule-based models", in Franz Hölker (Ed.), *Scales, Hierarchies and Emergent Properties in Ecological Models*, Frankfurt am Main, Peter Lang, 2002, 95–104.

Kurth (W.), Sloboda (B.), "Sensitive growth grammars specifying models of forest structure", Competition and Plant–Herbivore Interaction, Proceedings of the IUFRO 4. 11 Congress "Forest Biometry, Modelling and Information Science", Greenwich (UK), 25–29 June 2001, 15 pages.

Langton (C.G.) (Ed.), *Artificial Life*, Proceedings of an interdisciplinary workshop on the synthesis and simulation of living systems, September 1987, Santa Fe, New Mexico. Boston, Addison-Wesley, 1989.

Lassègue (J.), "Turing, l'ordinateur et la morphogenèse", *La Recherche*, 1998, No. 305, January, 76–77.

Lassègue (J.), *Turing*, Paris, Les Belles Lettres, 1998.

Latil (de) (P.), *La pensée artificielle: introduction à la cybernétique*, Paris, Gallimard, 1953.

Laubichler (M.D.), Müller (G.B.) (Eds), *Modeling Biology: Structures, Behaviors, Evolution*, Cambridge, The MIT Press, 2007.

Lecoustre (R.), Reffye (de) (P.), "AMAP, un modeleur de végétaux, un ensemble de logiciels de CAO/DAO à l'usage des professionnels de l'aménagement et des paysages", *Revue Horticole Suisse* [Swiss Horticulture Journal], 1993, Vol. 66, No. 6/7, 142–146.

Legay (J.M.), "Éléments d'une théorie générale de la croissance d'une population", *Bulletin of Mathematical Biophysics*, 1968, Vol. 30, No. 1, 33–46.

Legay (J.M.), "Contribution à l'étude de la forme des plantes: discussion d'un modèle de ramification", *Bulletin of Mathematical Biophysics*, 1971, Vol. 33, No. 3, 387–401.

Legay (J.M.), "La méthode des modèles, état actuel de la méthode expérimentale", *Informatique et Biosphère* [Computer Science and Biosphere], Paris, 1973, 5–73.

Legay (J.M.), *L'expérience et le modèle. Un discours sur la méthode*, Paris, INRA éditions, 1997.

Lenhard (J.), "Nanoscience and the Janus-faced character of simulations", in D. Baird, A. Nordmann, J. Schummer (Eds), *Discovering the Nanoscale*, Amsterdam, IOS Press, 2004, 93–100.

Lenhard, (J.), "Artificial, false, and performing well", in G. Gramelsberger (Ed.), *From Science to Computational Sciences*, Zürich, Diaphanes, 2011, 165–176.

Lenhard (J.), Winsberg (E.), "Holism, entrenchment, and the future of climate model pluralism", *Studies in History and Philosophy of Modern Physics*, 2010, Vol. 41, No. 3, 253–262.

Lénine (V.I.), *Matérialisme et empiriocriticisme*, Moscow, Éditions du progrès, 1908; trans.: Éditions sociales, Paris, 1973.

Leopold (L.B.), "Trees and streams: the efficiency of branching patterns", *Journal of Theoretical Biology*, 1971, Vol. 31, No. 2, 339–354.

Letort (V.), Mahe (P.), Cournede (P.H.), de Reffye (P.), Courtois (B.), "Quantitative genetics and functional-structural plant growth models: simulation of quantitative trait loci detection for model parameters and application to potential yield optimization", *Annals of Botany*, 2008, Vol. 101, No. 8, 1243–1254.

Lévy (P.), *La machine univers: création, cognition et culture informatique*, Paris, La Découverte, 1987; republished by Seuil – Point Sciences, 1992.

Lindenmayer (A.), "Life cycles as hierarchical relations", in J.R. Gregg, F.T.C. Harris (Eds), *Form and Strategy in Science: Studies Dedicated to Joseph Henry Woodger on the Occasion of his Seventieth Birthday*, Dordrecht, D. Reidel Publishing Company, 1964, 416–470.

Lindenmayer (A.), "Mathematical models for cellular interactions in development. I. filaments with one-sided inputs", *Journal of Theoretical Biology*, 1968, Vol. 18, No. 3, 280–299.

Lindenmayer (A.), "Mathematical models for cellular interactions in development. II. simple and branching filaments with two-sided inputs", *Journal of Theoretical Biology*, 1968, Vol. 18, No. 3, 300–315.

Lindenmayer (A.), "Developmental systems without cellular interactions, their languages and grammars", *Journal of Theoretical Biology*, 1971, Vol. 30, No. 3, 455–484.

Lindenmayer (A.), "Cellular automata, formal languages and developmental systems", in P. Suppes, L. Henkin, A. Joja, G.R.C. Moisil (Eds), *Logic, Methodology and Philosophy of Science IV, Proceedings of the 4th international Congress for Logic, Methodology and Philosophy of Science, Bucharest, 1971*, Amsterdam, North Holland Publishing Company, 1973, 677–691.

Lindenmayer (A.), Rozenberg (G.) (Eds), *Automata, Languages, Development*, Amsterdam, North-Holland Publishing Company, 1976.

Livet (P.), "Essai d'épistémologie de la simulation multi-agents en sciences sociales", in F. Amblard, D. Phan (Eds), *Modélisation et simulation multi-agents*, Paris, Hermes, 2006, 193–218.

Lotka (A.J.), *Elements of Physical Biology, 1924*; 2nd edition: *Elements of Mathematical Biology*, Dover Publications, New York, 1956.

Lotodé (R.), "Possibilités d'amélioration de l'expérimentation sur cacaoyers", *Café, Cacao, Thé*, 1971, Vol. 15, No. 2, April–June, 91–103.

Lück (H.B.), "Élementary behavioural rules as a foundation for morphogenesis", *Journal of Theoretical Biology*, 1975, Vol. 54, No. 1, 23–34.

Luquet (D.), "Suivi de l'état hydrique des plantes par infrarouge thermique", thèse de doctorat de l'Institut National Agronomique Paris-Grignon, 2002.

Mach (E.), *Analyse der Empfindungen*, Iéna, 1911, 1922; translation of the definitive posthumous edition of 1922: *L'analyse des sensations: le rapport du physique au psychique*, Paris, Éditions Jacqueline Chambon, 1996.

Mackenzie (D.A.), *Statistics in Britain, 1865–1960*, Edinburgh, Edinburgh University Press, 1981.

Malécot (G.), *Les mathématiques de l'hérédité*, Paris, Masson, 1948.

Malézieux (E.), Trébuil (G.), Jaeger (M.) (Eds), *Modélisation des agroécosystèmes et aide à la décision*, joint publication: Montpellier and Versailles, Librairie du CIRAD and INRA éditions, 2001.

Mandelbrot (B.), *The Fractal Geometry of Nature*, New York, Freeman, 1977.

Mayntz (R.), "Research technology, the computer, and scientific progress", in G. Gramelsberger (Ed.), *From Science to Computational Sciences*, Zürich, Diaphanes, 2011, 195–207.

McLeod (J.), "Computer modeling and simulation: the changing challenge", *Simulation*, 1986, Vol. 46, No. 3, March, 114–118.

Metropolis (N.), Ulam (S.), "A property of randomness of an arithmetical function", *American Mathematical Monthly*, 1953, Vol. 60, No. 4, 252–253.

Minsky (M.), "Matter, mind and models", in W.A. Kalenich (Ed.), *Proceedings of the International Federation for Information Processing (IFIP) Congress*, London, Macmillan, 1965, 45–49.

Mitchell (S.D.), *Biological Complexity and Integrative Pluralism*, Cambridge, Cambridge University Press, 2003.

Mondzain (M.J.), *Image, icône, économie*, Paris, Seuil, 1996.

Morange (M.), *Histoire de la biologie moléculaire*, Paris, La Découverte, 1994.

Morgan (M.S.), "Experiments vs. models: new phenomena, inference, and surprise", *Journal of Economic Methodology*, 2005, Vol. 12, No. 2, 317–329.

Morgan (M.S.), Morrison (M.) (Eds), *Models as Mediators*, Cambridge, Cambridge University Press, 1999.

Morrison (M.), *Reconstructing Reality. Models, Mathematics, and Simulations*, Oxford, Oxford University Press, 2015.

Moulines (C.U.), *La philosophie des sciences*, Paris, Éditions ENS/Rue d'Ulm, 2006.

Müller (J.P.) (Ed.), "Le statut épistémologique de la simulation", *Proceedings of the 10th "journées de Rochebrune": Interdisciplinary Meetings on Complex and Artificial Systems*, Paris, Éditions de l'École Nationale Supérieure de Télécommunications de Paris (ENST), 2003.

Murray (C.D.), "The physiological principle of minimum work applied to the angle of branching arteries", *The Journal of General Physiology*, 1926, Vol. 9, No. 6, July, 835–841.

Murray (C.D.), "A relationship between circumference and weight in trees and its bearing on branching angles", *The Journal of General Physiology*, 1927, Vol. 10, No. 5, May, 725–729.

Murray (C.D.), "The physiological principle of minimum work: a reply", *The Journal of General Physiology*, 1931, Vol. 14, No. 4, March, 445.

Nagel (E.), *The Structure of Science*, Indianapolis, Hackett Publishing Company, 1960.

Naylor (T.H.), Balintfy (J.L.), Burdick (D.S.), Chu (K.), *Computer Simulation Techniques*, New York, John Wiley and Sons, 1966.

Neumann (von) (J.), "The role of the digital procedure in reducing the noise level" (1948), in A.H. Taub (Ed.), *John von Neumann – Collected Works*, Vol. V, London, Pergamon Press, 1962.

Neumann (von) (J.), "The general and logical theory of automata", in Lloyd A.J. (Ed.), *Cerebral Mechanisms and Behaviour*, New York, John Wiley and Sons, 1951; trans. J.P. Auffrand: *Théorie générale et logique des automates*, Paris, Champ-Vallon, 1996.

Neumann (von) (J.), *The Computer and the Brain*, New Haven, Yale University Press, 1958.

Neumann (von) (J.), *Theory of Self-Reproducing Automata*, edited and completed by A.W. Burks, Urbana, University of Illinois Press, 1966.

Niklas (K.J.), "Computer simulations of branching-patterns and their implications on the evolution of plants", in L.J. Gross, R.M. Miura (Eds) *Some Mathematical Questions in Biology, Lecture on Mathematics in the Life Sciences*, Vol. 18, American Mathematical Society, Providence, Rhode Island, 1986, 1–50.

Niklas (K.J.), "The evolution of plant body plans: a biomechanical perspective", *Annals of Botany*, 2000, Vol. 85, No. 4, 411–438.

Niklas (K.J.), Spatz (H.C.), "Growth and hydraulic (not mechanical) constraints govern the scaling of tree height and mass", *PNAS*, 2004, Vol. 101, No. 44, 15661–15663.

Nouvel (P.) (Ed.), *Enquête sur le concept de modèle*, Paris, PUF, 2002.

Odum (H.T.), Odum (E.C.), *Modelling for all Scales: An Introduction to System Simulation*, San Diego, Academic Press, 2000.

Oldeman (R.A.A.), "L'architecture de la forêt guyanaise", State Doctoral thesis, Université de Montpellier II, 1972.

Oldeman (R.A.A), *L'architecture de la forêt guyanaise*, Paris, Mémoire de l'ORSTOM, No. 73, 1974.

Parker (W.), "Does matter really matter: computer simulations, experiments and materiality", *Synthese*, 2009, Vol. 169, No. 3, 483–496.

Parrochia (D.), "L'expérience dans les sciences: modèles et simulation", in *Qu'est-ce que la vie? Université de tous les savoirs [University of all knowledge]*, Paris, Odile Jacob, 2000, 193–203.

Parvais (J.P.), Reffye (de) (P.), Lucas (P.), "Observations sur la pollinisation libre chez *Theobroma Cacao*: analyse mathématique des données et modélisation", *Café, Cacao, Thé*, 1977, Vol. 21, No. 4, October–December, 253–262.

Pavé (A.), *Modélisation en biologie et en écologie*, Lyon, Aléas Éditeur, 1994.

Pavé (A.) et al., *Première revue externe de l'unité de modélisation des plantes*, Montpellier, CIRAD, 1996.

Pearson (K.), *The Grammar of Science*, London, The Temple Press, 1892.

Perttunen (J.), Nikinmaa (E.), Lechowicz (M.J), Sievänen (R.), Messier (C.), "Application of the functional-structural tree model LIGNUM to sugar maple saplings (*Acer saccharum* Marsh) growing in forest gaps", *Annals of Botany*, 2001, Vol. 88, No. 3, 471–481.

Perttunen (J.), Sievänen (R.), Nikinmaa (E.), Salminen (H.), Saarenmaa (H.), Väkeva (J.), "LIGNUM: a tree model based on simple structural units", *Annals of Botany*, 1996, Vol. 77, No. 1, 87–98.

Petitot (J.), *Morphogenèse du sens – I*, Paris, PUF, 1985.

Petitot (J.) (Ed.), *Logos et théorie des catastrophes*, Genève, Éditions Patino, coll. Colloques de Cerisy, 1988.

Phan (D.), Amblard (F.), *Agent-based Modelling and Simulation in the Social and Human Sciences*, Oxford, The Bardwell Press, 2007.

Phan (D.), Varenne (F.), "Épistémologie dans une coquille de noix: concevoir et expérimenter", in F. Amblard, D. Phan (Eds), *Modélisation et Simulation Multi-agents: Applications aux Sciences de l'Homme et de la Société*, London, Hermes, 2006, 104–119.

Phan (D.), Varenne (F.), "Agent-based models and simulations in economics and social sciences: from conceptual exploration to distinct ways of experimenting", *Journal of Artificial Societies and Social Simulation*, 2010, Vol. 13, No. 1, 5, http://jasss.soc. surrey.ac.uk/13/1/5.html.

Piaget (J.), "La représentation 'concrète'", in J. Piaget (Ed.), *Logique et connaissance scientifique*, Paris, Pléiade, 1967, 772–778.

Picard (J.F.), *La république des savants: la recherche française et le CNRS*, Paris, Flammarion, 1990.

Pichot (A.), *Histoire de la notion de vie*, Paris, Gallimard-TEL, 1993.

Pinel (E.), *Les fondements de la biologie mathématique non statistique*, Paris, Maloine, 1973.

Poincaré (H.), *La valeur de la science*, Paris, 1905; republication: Champs-Flammarion, 1994.

Polya (G.), *How to Solve It*, Princeton, Princeton University Press, 1945; 2nd edition: 1957.

Pouget (J.M.), *La science goethéenne des vivants: de l'histoire naturelle à la biologie évolutionniste*, Bern, Peter Lang, 2001.

Prenant (M.), *Biologie et marxisme*, Paris, Éditions Sociales Internationales, 1935.

Prusinkiewicz (P.), "Modeling of spatial structure and development of plants: a review", *Scientia Horticulturae*, 1998, Vol. 74, Nos 1–2, 113–149.

Prusinkiewicz (P.), "Modeling plant growth and development", *Current Opinion in Plant Biology*, 2003, Vol. 7, No. 1, 1–5.

Prusinkiewicz (P.), Lindenmayer (A.), *The Algorithmic Beauty of Plants*, New York, Springer Verlag, 1990.

Prusinkiewicz (P.), Lindenmayer (A.), Hanan (J.), "Developmental models of herbaceous plants for computer imagery purposes", *Computer Graphics*, 1988, Vol. 22, No. 4, 141–150.

Ramunni (G.), *La physique du calcul: histoire de l'ordinateur*, Paris, Hachette, 1989.

Rapaport (D.C.), *The Art of Molecular Dynamics Simulation*, Cambridge, Cambridge University Press, 1995.

Rashevsky (N.), "Foundations of mathematical biophysics", *Philosophy of Science*, 1934, Vol. 1, No. 2, 176–196.

Rashevsky (N.), *Mathematical Biophysics* (one volume), Chicago, University of Chicago Press, 1938; 2nd edition (two volumes): 1948.

Rashevsky (N.), "A contribution to the search of general mathematical principles in biology", *Bulletin of Mathematical Biophysics*, 1958, Vol. 20, No. 1, 71–93.

Rashevsky (N.), *Mathematical Biophysics: Physico-Mathematical Foundations of Biology* (two volumes), Chicago, University of Chicago Press, 1960.

Rashevsky (N.), *Mathematical Principles in Biology and their Applications*, Illinois, Charles C. Thomas Publisher, 1961.

Reffye (de) (P.), "La recherche de l'optimum en amélioration des plantes et son application à une descendance F1 de *Coffea arabusta*", *Café, Cacao, Thé*, 1974, Vol. 18, No. 3, 167–178.

Reffye (de) (P.), "Le contrôle de la fructification et de ses anomalies chez les *Coffea arabica, robusta* et leurs hybrides Arabusta", *Café, Cacao, Thé*, 1974, Vol. 18, No. 4, 237–254.

Reffye (de) (P.), "Formulation mathématique des facteurs de la fertilité dans le genre *coffea*", Third-cycle Doctoral thesis, Université Paris-Sud Orsay, 1975.

Reffye (de) (P.), "Modélisation et simulation de la verse du caféier, à l'aide de la théorie de la résistance des matériaux", *Café, Cacao, Thé*, 1976, Vol. 20, No. 4, 251–272.

Reffye (de) (P.), "Modélisation de l'architecture des arbres par des processus stochastiques. Simulation spatiale des modèles tropicaux sous l'effet de la pesanteur. Application au *Coffea Robusta*", State Doctoral thesis, Université Paris-Sud, Orsay, 1979.

Reffye (de) (P.), "Modèle mathématique aléatoire et simulation de la croissance et de l'architecture du caféier *Robusta*. 1ère partie. Étude du fonctionnement des méristèmes et de la croissance des axes végétatifs", *Café, Cacao, Thé*, 1981, Vol. 25, No. 2, 83–104.

Reffye (de) (P.), "Modèle mathématique aléatoire et simulation de la croissance et de l'architecture du caféier *Robusta*. 2ème partie. Étude de la mortalité des méristèmes plagiotropes", *Café, Cacao, Thé*, 1981, Vol. 25, No. 3, 219–230.

Reffye (de) (P.), "Modèle mathématique aléatoire et simulation de la croissance et de l'architecture du caféier *Robusta*. 3ème partie. Étude de la ramification sylleptique des rameaux primaires et de la ramification proleptique des rameaux secondaires", *Café, Cacao, Thé*, 1982, Vol. 26, No. 2, 77–96.

Reffye (de) (P.), "Modèle mathématique aléatoire et simulation de la croissance et de l'architecture du caféier *Robusta*. 4ème partie. Programmation sur micro-ordinateur du tracé en trois dimensions de l'architecture d'un arbre", *Café, Cacao, Thé*, 1983, Vol. 27, No. 1, 3–20.

Reffye (de) (P.) *et al.* (Eds), *Document préparatoire à la revue externe de l'unité de modélisation des plantes*, CIRAD, Montpellier, 1996.

Reffye (de) (P.), Blaise (F.), Chemouny (S.), Jaffuel (S.), Fourcaud (T.), "Calibration of a hydraulic-based growth model of cotton plant", *Agronomie*, 1999, Vol. 19, No. 3/4, 265–280.

Reffye (de) (P.), Blaise (F.), Fourcaud (T.), Houllier (F.), Barthélémy (D.), "Un modèle écophysiologique de la croissance et de l'architecture des arbres et de leurs interactions", in B. Andrieu (Ed.), *Modélisation architecturale*, Proceedings of the seminar of 10–12 March 1997, Bioclimatology Department, INRA-Grignon, 129–135.

Reffye (de) (P.), Dinouard (P.), Barthélémy (D.), "Modélisation et simulation de l'architecture de l'orme du Japon *Zelkova serrata* (Thunb.) Makino *(Ulmaceae)*: la notion d'axe de référence", in C. Edelin (Ed.), 2ème Colloque International sur l'Arbre [2nd International Symposium on the Tree], 10–15 September 1990, Montpellier, Naturalia Monspeliensa, Special Edition A7, 251–266.

Reffye (de) (P.), Edelin (C.), Françon (J.), Jaeger (M.), Puech (C.), "Plants models faithful to botanical structure and development", *Computer Graphics*, 1988, Vol. 22, No. 4, 151–158.

Reffye (de) (P.), Edelin (C.), Jaeger (M.), "Modélisation de la croissance des plantes", *La Recherche*, 1989, Vol. 20, No. 207, 158–168.

Reffye (de) (P.), Elguero (E.), Costes (E.), "Growth units construction in trees: a stochastic approach", *Acta Biotheoretica*, 1991, Vol. 39, Nos 3–4, 325–342.

Reffye (de) (P.), Goursat (M.), Quadrat (J.P.), Hu (B.G.), "The dynamic equations of the tree morphogenesis GreenLab Model", in B. G. Hu, M. Jaeger (Eds), *Plant Growth Modeling and Applications, Proceedings of 2003 International Symposium on Plant Growth Modeling, Simulation, Visualization and their Applications (PMA03)*, Beijing, China, Tsinghua University Press, 2003, 108–116.

Reffye, (de) (P.), Jaeger (M.), Barthélémy (D.), Houllier (F.) (Eds), *Architecture et croissance des plantes. Modélisation et application*, Paris, Quae, 2016.

Reffye (de) (P.), Snoeck (J.), "Modèle mathématique de base pour l'étude et la simulation de la croissance et de l'architecture du *Coffea Robusta*", *Café, Cacao, Thé*, 1976, Vol. 20, No. 1, 11–31.

Roger (J.), *Pour une histoire des sciences à part entière*, Paris, Albin Michel, 1995.

Rohrlich (F.), "Computer simulation in the physical sciences", *PSA (Philosophy of Science Association)*, 1990, Vol. 2, 507–518.

Roll-Hansen (N.), "E.S. Russell and J.H. Woodger: the failure of two twentieth-century opponents of mechanistic biology", *Journal of the History of Biology*, 1984, Vol. 17, No. 3, 399–428.

Rosen (R.), "A relational theory of biological systems", *Bulletin of Mathematical Biophysics*, 1958, Vol. 20, No. 3, 245–260.

Rosen (R.), "The representation of biological systems from the standpoint of the theory of categories", *Bulletin of Mathematical Biophysics*, 1958, Vol. 20, No. 4, 317–341.

Rosen (R.), "Turing's morphogens, two-factor systems and active transport", *Bulletin of Mathematical Biophysics*, 1968, Vol. 30, No. 3, 493–499.

Rosen (R.), *Essays on Life Itself*, New York, Columbia University Press, 2000.

Rosenblueth (A.), Wiener (N.), "The role of models in science", *Philosophy of Science*, 1945, Vol. 12, No. 4, October, 316–321.

Sachs (T.), "Consequences of the inherent developmental plasticity of organ and tissue relations", *Evolutionary Ecology*, 2002, Vol. 16, No. 3, 243–265.

Sachs (T.), Novoplansky (A.), "Tree form: architectural models do not suffice", *Israel Journal of Plant Sciences*, 1995, Vol. 43, No. 3, 203–212.

Saint-Sernin (B.), *La raison au XXème siècle*, Paris, Seuil, 1995.

Sauvan (J.) "Méthode des modèles et connaissance analogique", *Revue d'Agressologie*, 1966, Vol. 7, No. 1, 9–18.

Schmid (A.F.), *L'âge de l'épistémologie*, Paris, Kimè, 1998.

Schmitt (S.), *Histoire d'une question anatomique: la répétition des parties*, Paris, Éditions du Muséum National d'Histoire Naturelle, 2004.

Schreider (E.), *La biométrie*, Paris, PUF, 1960; 2nd edition: 1967.

Schruben (L.W.), "Establishing the credibility of simulations", *Simulation*, 1980, Vol. 34, No. 3, March, 101–105.

SCS Technical Committees, "Terminology for model credibility", *Simulation*, 1979, Vol. 32, No. 3, March, 103–104.

Segal (J.), *Le zéro et le un: histoire de la notion scientifique d'information au XXème siècle*, Paris, Syllepse, 2003.

Shannon (R.E.), "Introduction to the art and science of simulation", in D.J. Medeiros, E.F. Watson, J.S. Carson, M.S. Manivannan (Eds), *Proceedings of the 1998 Winter Simulation Conference*, 13 December, Washington, USA. New York, IEEE, 7–14.

Silk (W.K.), Erickson (R.O.), "Kinematics of plant growth", *Journal of Theoretical Biology*, 1979, Vol. 76, No. 4, 481–501.

Sinaceur (H.), *Corps et modèles*, Paris, Vrin, 1991; 2nd edition, corrected: 1999.

Sismondo (S.), "Simulation as a new style of research", in G. Gramelsberger (Ed.), *From Science to Computational Sciences*, 2011, Zürich, Diaphanes, 151–163.

Skellam (J.G.), "Some philosophical aspects of mathematical modelling in empirical science with special reference to ecology", in J.N.R. Jeffers (Ed.), *Mathematical Models in Ecology*, Oxford, London, Blackwell Scientific, 1977, 27.

Slobodkin (L.B.), "Meta-models in theoretical ecology", *Ecology*, 1958, Vol. 39, 3.

Smith (A.R.), "Plants, fractals, and formal languages", *Computer Graphics*, 1984, Vol. 18, No. 3, July, 1–10.

Soler (C.), Sillion (F.), Blaise (F.), Reffye (de) (P.), *A Physiological Plant Growth Simulation Engine based on Accurate Radiant Energy Transfer*, Le Chesnay, INRIA research report, No. 4116, February 2001.

Sommerhoff (G.), *Analytical Biology*, Oxford, Oxford University Press, 1950.

Sorensen (R.A.), *Thought Experiments*, Oxford, Oxford University Press, 1992.

Stahl (W.R.), "The role of models in theoretical biology", *Progress in Theoretical Biology*, 1967, Vol. 1, No. 1, 165–218.

Stöckler (M.), "On modeling and simulations as instruments for the study of complex systems", in M. Carrier, G.J. Massey, L. Ruetsche (Eds), *Science at Century's End*, Pittsburgh, University of Pittsburgh Press, 2000, 355–373.

Suppes (P.), "A comparison of the meaning and the uses of models in mathematics and the empirical sciences", in Hans Freudenthal (Ed.), *The Concept and the Role of the Model in Mathematics and Natural and Social Sciences*, Dordrecht, D. Reidel Publishing Company, 1961, 163–177.

Szilard (A.L.), Quinton (R.E.), "An interpretation for *DOL* systems by computer graphics", *The Science Terrapin*, 1979, Vol. 4, 8–13.

Tannier (C.), "Analyse et simulation de la concentration et de la dispersion des implantations humaines de l'échelle micro-locale à l'échelle régionale: modèles multi-échelles et trans-échelles", Habilitation Doctorate, Besançon, University of Burgundy, 2017.

Taton (R.) (Ed.), *La science contemporaine, Vol. 1: le XIXème siècle*, Paris, PUF, 1961; édition Quadrige: 1995.

Taton (R.) (Ed.), *La science contemporaine, Vol. 2: le XXème siècle – années 1900–1960*, Paris, PUF, 1964; édition Quadrige: 1995.

Teissier (G.), "La description mathématique des faits biologiques", *Revue de métaphysique et de morale*, 1936, Vol. 43, No. 1, 55–87.

Teissier (G.), "Les lois quantitatives de la croissance", Paris, *Rapport de l'Association des Physiologistes [Report of the Association of Physiologists]*, collection of presentations on biometry and biological statistics, 455, XI, 1937.

Teissier (G.), *Matérialisme dialectique et biologie*, Paris, Éditions Sociales, 1946.

Thom (R.), *Modèles mathématiques de la morphogenèse*, Paris, UGE-10/18, 1974.

Thom (R.), *Stabilité structurelle et morphogenèse*, New York, W.A. Benjamin Inc., 1972; revised and supplemented second edition: Paris, InterEditions, 1977.

Thompson, d'Arcy (Sir), *On Growth and Form*, Cambridge, Cambridge University Press, 1917; new edition, 1942; abridged in 1961 by J.T. Bonner; trans. by D. Teyssié: *Forme et croissance*, Paris, Seuil, 1994.

Tomassone (R.), Dervin (C.), Masson (J.P.), *Biométrie. Modélisation de phénomènes biologiques*, Paris, Masson, 1993.

Treuil (J.P.), Mullon (C.), "Expérimentation sur mondes artificiels: pour une réflexion méthodologique", in Blasco (F.) (Ed.), *Tendances nouvelles en modélisation pour l'environnement, Proceedings of symposium of 16–17 January 1996 – Programme Environnement du CNRS [CNRS Environment Programme]*, Paris, éditions Elsevier, 1997, 425–431.

Turing (A.M.), "The chemical basis of morphogenesis", *Philosophical Transactions of the Royal Society, B*, 1952, Vol. 237, No. 641, 37–72.

Turing (A.M.), Girard (J.Y.), *La machine de Turing*, Paris, Seuil, 1995.

Ulam (S.), "On the Monte-Carlo method", *Proceedings of the Second Symposium on Large-Scale Digital Calculating Machinery, 1949*, Cambridge, Harvard University Press, 1951, 207–212.

Ulam (S.), "Infinite models in physics", in *Applied Probability. Proceedings of Symposia in Applied Mathematics*, Vol. VII, New York, Toronto and London, McGraw-Hill, for the American Mathematical Society, Providence, RI, 1957, 87–95.

Ulam (S.), "On some mathematical problems connected with patterns of growth of figures", *Proceedings of Symposia in Applied Mathematics*, American Mathematical Society, 1962, Vol. 14, 215–224.

Ulam (S.), "Patterns of growth of figures: mathematical aspects", in G. Kepes (Ed.), *Module, Proportion, Symmetry, Rhythm*, New York, Braziller, 1966, 64–74.

Ulam (S.), *Adventures of a Mathematician*, Berkeley, University of California Press, 1976; supplemented republication: 1991.

Van Fraassen (B.), *The Scientific Image*, Oxford, Clarendon Press, 1980.

Varenne (F.), "What does a computer simulation prove?", in N. Giambiasi, C. Frydman (Eds), *Simulation in Industry*, Proceedings of the 13th European Simulation Symposium, Marseille, 18–20 October, SCS Europe Bvba, Ghent, 2001, 549–554.

Varenne (F.), "La simulation conçue comme expérience concrète", in J.P. Müller (Ed.), *Le statut épistémologique de la simulation, Proceedings of the 10th journées de Rochebrune*, Paris, Éditions de l'École Nationale Supérieure de Télécommunications de Paris (ENST), 2003, 299–313.

Varenne (F.), "La simulation informatique face à la méthode des modèles", *Natures Sciences Sociétés*, 2003, Vol. 11, No. 1, 16–28.

Varenne (F.), "Le destin des formalismes: à propos de la forme des plantes: pratiques et épistémologies des modèles face à l'ordinateur", PhD, University of Lyon, 2004, https://tel.archives-ouvertes.fr/tel-00008810.

Varenne (F.), "Un aperçu sur la biologie théorique au XXème siècle", *Traces de Futurs 2004*, Proceedings of the XVIth Carcassonne Interdisciplinary Symposium organized by Université Paul Sabatier (Toulouse) and ADREUC (Association pour le Développement des Rencontres et des Échanges Universitaires et Culturels), Arques, 2005, 34–45.

Varenne (F.), "Bachelard avec la simulation informatique: nous faut-il reconduire sa critique de l'intuition?", in R. Damien, B. Hufschmitt (Eds), *Bachelard: confiance raisonnée et défiance rationnelle*, Besançon, Presses Universitaires de Franche-Comté, 2006, 111–143.

Varenne (F.), "La simulation d'objets complexes: retour à un 'sens commun' simulé?", *Natures Sciences Sociétés*, 2006, Vol. 14, No. 1, 63–65.

Varenne (F.), "Nicholas Rashevsky (1899–1972): de la biophysique à la biotopologie", *Cahiers d'Histoire et de Philosophie des Sciences*, special edition: Actes du Congrès National d'Histoire des Sciences et des Techniques de Poitiers (May 2004), 2006, 162–163.

Varenne (F.), "Optimalité et morphogenèse: le cas des plantes au XXème siècle", *Bulletin d'Histoire et d'Épistémologie des Sciences de la Vie*, 2006, Vol. 13, No. 1, éditions Kimé, 89–117.

Varenne (F.), *Les notions de métaphore et d'analogie dans les épistémologies des modèles et des simulations*, preface by A.F. Schmid, Paris, Éditions Pétra, 2006.

Varenne (F.), *Du modèle à la simulation informatique*, Paris, Vrin, 2007.

Varenne (F.), "Fragmenter les modèles: simulation numérique et simulation informatique", in P.A. Miquel (Ed.), *Biologie du XXIème siècle: evolution des concepts fondateurs*, Brussels, De Boeck, 2008, Chapter 11, 265–295.

Varenne (F.), *Qu'est-ce que l'informatique?*, Paris, Vrin, 2009.

Varenne (F.), "Framework for models & simulations with agents in regard to agent simulations in social sciences: emulation and simulation", in A. Muzy, D. Hill and B. Zeigler (Eds), *Modeling & Simulation of Evolutionary Agents in Virtual Worlds*, Clermont-Ferrand, Presses Universitaires Blaise Pascal, 2010, 53–84.

Varenne (F.), "Les simulations computationnelles dans les sciences sociales", *Nouvelles Perspectives en Sciences Sociales*, 2010, Vol. 5, No. 2, 17–49, www.erudit.org/revue/npss/2010/v5/n2/index.html.

Varenne (F.), *Formaliser le vivant: lois, théories, modèles?*, Paris, Hermann, 2010.

Varenne (F.), *Modéliser le social: méthodes fondatrices et évolutions récentes*, Paris, Dunod, 2011.

Varenne (F.), "La reconstruction phénoménologique par simulation: vers une épaisseur du *simulat*", in D. Parrochia, V. Tirloni, *Formes, systèmes et milieux techniques après Simondon*, Lyon, Jacques André Éditeur, 2012, 107–123.

Varenne (F.), *Théorie, réalité, modèle*, Paris, Matériologiques, 2012.

Varenne (F.), *Chains of Reference in Computer Simulations*, working paper selected and published by FMSH, FMSH-WP-2013–51, 2013, 32 pages, https://halshs.archives-ouvertes.fr/halshs-00870463.

Varenne (F.), *Théories et modèles en sciences humaines. Le cas de la géographie*, Paris, Éditions Matériologiques, 2017.

Varenne (F.), Chaigneau (P.), Petitot (J.), Doursat (R.), "Programming the emergence in morphogenetically architected complex systems", *Acta Biotheoretica*, 2015, Vol. 63, No. 3, September, 295–308.

Varenne (F.), Silberstein (M.) (Eds), *Modéliser & simuler: épistémologies et pratiques de la modélisation et de la simulation*, Vol. 1, Éditions Matériologiques, Paris, 2013.

Varenne (F.), Silberstein (M.), Dutreuil (S.), Huneman (P.), (Eds), *Modéliser & simuler: épistémologies et pratiques de la modélisation et de la simulation, Vol. 2*, Paris, Éditions Matériologiques, 2014.

Vessereau (A.), *La statistique*, Paris, PUF; QSJ, 1ère édition: 1947; 18ème édition: 1992.

Vessereau (A.), *Méthodes statistiques en biologie et agronomie*, first edition: Paris, J.B. Baillière 1948; second edition: Paris, J.B. Baillière, 1960; reprint: Paris, Lavoisier, 1988.

Vinci (da) (L.), *The Notebooks of Leonardo da Vinci*, English translation by Edward MacCurdy, New York, Reynal and Hitchcock, 1939; reprint: New York, Braziller, 1955.

Volterra (V.), *Leçons sur la théorie mathématique de la lutte pour la vie*, Paris, Gauthier-Villars, 1931; reprint: Paris, Éditions Jacques Gabay, 1990.

Waddington (C.H.), *New Patterns in Genetics and Development*, New York, Columbia University Press, 1962.

Wagensberg (J.), *Ideas sobre la complejidad del mundo*, Barcelona, Tusquets Editores, 1985; trans.: *L'âme de la méduse: idées sur la complexité du monde*, Paris, Seuil, 1997.

Walliser (B.), *Systèmes et modèles: introduction critique à l'analyse des systèmes*, Paris, Seuil, 1977.

Wardlaw (C.W.), *Essays on Form in Plants*, Manchester, Manchester University Press, 1968.

Waterman (T.H.), Morowitz (H.J.), *Theoretical and Mathematical Biology*, New York, Toronto and London, Blaisdell Publishing Company, 1965.

White (J.), "The plant as a metapopulation", *Annual Review of Ecology and Systematics*, 1979, Vol. 10, 109–145.

Whitehead (A.N.), Russell (B.), *Principia Mathematica to 56*, London, Cambridge University Press, 1910; abridged edition: 1962; reprint: 1970.

Wiener (N.), *Cybernetics, or Control and Communications in the Animal and the Machine*, New York, John Wiley, 1948.

Wimsatt (W.C.), *Re-Engineering Philosophy for Limited Beings*, Cambridge, Harvard University Press, 2007.

Winsberg (E.), "Sanctioning models: the epistemology of simulation", *Science in Context*, 1999, Vol. 12, No. 2, 275–292.

Winsberg (E.), "Simulated experiments: methodology for a virtual world", *Philosophy of Science*, 2003, Vol. 70, No. 1, 105–125.

Winsberg (E.), "Handshaking your way to the top: simulation at the nanoscale", *Philosophy of Science*, 2006, Vol. 73, No. 5, 582–594.

Winsberg (E.), "A tale of two methods", *Synthese*, 2009, Vol. 169, No. 3, 575–592.

Winsberg (E.), *Science in the Age of Computer Simulation*, Chicago, University of Chicago Press, 2010.

Wolfram (S.), *A New Kind of Science*, Champaign, Wolfram Media Inc., 2002.

Woodger (J.H.), *The Axiomatic Method in Biology*, Cambridge, Cambridge University Press, 1937.

Xu (L.), Henke (M.), Zhu (J.), Kurth (W.), Buck-Sorlin (G.H.), "A functional-structural model of rice linking quantitative genetic information with morphological development and physiological processes", *Annals of Botany*, 2011, Vol. 107, No. 5, 817–828.

Yan (H.P.), Reffye (de) (P.), Le Roux (J.), Hu (B.G.), "Study of plant growth behaviors simulated by the functional-structural plant model GreenLab", in B. G. Hu, M. Jaeger (Eds), *Plant Growth Modeling and Applications, Proceedings of 2003 International Symposium on Plant Growth Modeling, Simulation, Visualization and their Applications (PMA03)*, Beijing, China, Tsinghua University Press, October 2003, 118–125.

Yates, (F.), *The Art of Memory*, London, Routledge, 1966.

Zamir (M.), "The role of shear forces in arterial branching", *The Journal of General Physiology*, 1976, Vol. 67, 213–222.

Zeigler (B.P.), Praehofer (H.), Kim (T.G.), *Theory of Modeling and Simulation: Integrating Discrete Event and Continuous Complex Dynamic Systems*, New York, Academic Press, 2000.

Zhan (Z.G.), Reffye (de) (P.), Houllier (F.), Bao-Gang (H.), "Fitting a structural-functional model with plant architectural data", in B. G. Hu, M. Jaeger (Eds), *Plant Growth Modeling and Applications, Proceedings of 2003 International Symposium on Plant Growth Modeling, Simulation, Visualization and their Applications (PMA03)*, Beijing, China, Tsinghua University Press, October 2003, 236–243.

Zhao (X.), Reffye (de) (P.), Barthélémy (D.), Bao-Gang (H.), "Interactive simulation of plant architecture based on a dual-scale automaton model", in B. G. Hu, M. Jaeger (Eds), *Plant Growth Modeling and Applications, Proceedings of 2003 International Symposium on Plant Growth Modeling, Simulation, Visualization and their Applications (PMA03)*, Beijing, China, Tsinghua University Press, October 2003, 144–153.

Index of names

Index of subjects